# Qualitative Methodologies for Geographers

## Issues and debates

*Edited by*

**Melanie Limb**

Senior Lecturer in Geography, University College Northampton

and

**Claire Dwyer**

Lecturer in Geography, University College London

A member of the Hodder Headline Group
LONDON
Co-published in the United States of America by
Oxford University Press Inc., New York

First published in Great Britain in 2001 by
Arnold, a member of the Hodder Headline Group
338 Euston Road, London NW1 3BH

**http://www.arnoldpublishers.com**

Co-published in the United States of America by
Oxford University Press Inc.,
198 Madison Avenue, New York, NY10016

*British Library Cataloguing in Publication Data*
A catalogue record for this book is available from the British Library

*Library of Congress Cataloging-in-Publication Data*
A catalog record for this book is available from the Library of Congress

ISBN 0 340 74225 9 (hb)
ISBN 0 340 74226 7 (pb)

1 2 3 4 5 6 7 8 9 10

Production Editor: Rada Radojicic
Production Controller: Iain McWilliams
Cover Design: Terry Griffiths

Typeset in 10/14pt Gill Light by Phoenix Photosetting, Chatham, Kent
Printed and bound in Great Britain by MPG Books Ltd, Bodmin, Cornwall

What do you think about this book? Or any other Arnold title?
Please send your comments to feedback.arnold@hodder.co.uk

# Qualitative
# Methodologies
# for Geographers

# Contents

# PART 6
## Writing

# PART 7
## Vignettes

# List of figures

# List of boxes

# Notes on contributors

**Annabelle Aish** graduated in geography from University College London in 2000 and is currently doing a Masters degree.

**Stuart Aitken** is Professor of Geography at San Diego State University. His books include *Family Fantasies and Community Space* (Rutgers University Press, 1998), *Place, Space, Situation and Spectacle: A Geography of Film* (co-editor with Leo Zonn, Rowman and Littlefield, 1994) and *Putting Children in Their Place* (Association of American Geographers, 1994). His current project is a monograph entitled *Under Moral Assault: Geographies of Kids and Teens* (under contract with Routledge). He has also published widely in academic journals and edited collections on qualitative methods, film, critical social theory, children, families and communities.

**Tracey Bedford** is a Research Associate with the Environment and Society Research Unit based in the Department of Geography at University College London. She has worked on a series of projects concerned with issues of sustainability and environmental responsibility, using qualitative and participatory methodologies. She is currently researching the relationships between participatory processes in the planning system and environmental practice.

**Katy Bennett** is a Research Fellow in the Centre for Rural Economy at the University of Newcastle upon Tyne. Her research interests focus on the internal dynamics of households and communities and how these influence responses to processes of restructuring. She is co-author (with H. Benyon and R. Hudson) of *Coalfields Regeneration: Dealing with the Consequences of Industrial Decline* (Policy Press, 2000).

**Jacquelin Burgess** is Professor of Geography at University College London and head of the Environment and Society Research Unit. She has been interested in methodological concerns for much of her career, and was one of the primary geographers to incorporate psychotherapeutic methods, such as in-depth groups, into qualitative research. She has worked on a variety of research projects which have involved focus groups and in-depth groups, particularly in the field of environmental attitudes and values.

**Ruth Butler** is a Lecturer in Applied Social Research at the University of Hull where she teaches courses on research methods and geographies of disability. Her research interests include geographies of disability, youth and sexuality. She is currently working with Gill Valentine and Tracey Skelton on an ESRC-funded project on the experiences of deaf, and lesbian and gay youths. Her work has been published in *Transactions of the Institute of British Geographers, Environment and Planning* D and the *Journal of Gender Studies*. She is co-editor of *Mind and Body Spaces: geographies of illness, impairment and disability* (Routledge, 1999).

**Philippa Capps** graduated in geography from Kings College London in 2000 and is now working for an IT company.

**Sarah Corke** is completing a geography degree at University College London and will graduate in 2001.

**Mike Crang** is a Lecturer in geography at the University of Durham. He has worked with a variety of qualitative methods on issues around heritage, memory, identity and tourism. He has written a text book, *Cultural Geography*, and edited *Thinking Space* (with Nigel Thrift), *Virtual Geographies* (with Phil Crang and Jon May), and a forthcoming collection, *Tourism: between place and performance* (with Simon Coleman).

**Lorraine Dowler** is an Assistant Professor of Geography at Penn State University. Her interests focus on the area of gender and nationalism. Her previous research has focused on issues of identity politics in Northern Ireland. She recently started an investigation of women in firefighting.

**Claire Dwyer** is a Lecturer in geography at University College London where she teaches courses on social and cultural geography, and contributes to courses on qualitative methods at both undergraduate and masters level. Her research interests are in gender, ethnicity and the cultural politics of multiculturalism. She is currently undertaking research on the transnationalities of British-Asian commodity culture. She was a member of the writing collective for *Feminist Geographies* (Longman, 1997) and co-author of *Geographies of New Femininities* (Longman, 1999.)

**Peter Jackson** is Professor of Human Geography at the University of Sheffield. His recent research has focused on the relationship between consumption and identity, including ESRC-funded studies of Brent Cross and Wood Green shopping centres, the readership of men's lifestyle

magazines and the nature of transnational commodity culture. Recent
publications include *Shopping, place and identity* (Routledge, 1998),
*Commercial cultures* (Berg, 2000) and *Making sense of men's magazines*
(Polity, in press).

**James Kneale's** research centres on geographies of popular culture. He
is particularly interested in the circulation and consumption of everyday
texts and objects, and has written about the spaces of media audiences
and of drinking. He is currently editing a book on the geographies of
science fiction with Rob Kitchin at the University of Ireland, Maynooth.
James is a Lecturer in Cultural Geography at University College London.

**Audrey Kobayashi** is Professor of Geography at Queen's University,
Kingston, Canada. She has published extensively in the areas of anti-racism,
immigration, gender and human rights issues. Her latest major publication,
with Abigail Bakan, is *Employment Equity Policy in Canada: An Interprovincial
Comparison.* She is currently engaged in major research projects on
Japanese and Chinese immigrants to Canada.

**David Ley** is Professor of Geography at the University of British
Columbia. Since his doctoral dissertation he has been committed to the
application of interpretative method in human geography and to the
integration of qualitative and quantitative perspectives. Following a book
on *Gentrification, the New Middle Class and the Remaking of the Central City*
(1996) his research has been concerned with the social geographies of
immigration and ethnicity in gateway cities.

**Melanie Limb** is Senior Lecturer in Geography and Assistant Director of
the Centre for Children and Youth at University College Northampton.
She teaches social and cultural geography, qualitative methods and
philosophy. Her research interests are in the geographies of children and
young people, looking in particular at children's use of and values for space
and place and youth participation in local political processes.

**Robina Mohammad** conducted her doctoral research on political
geographies of gender in Spain and held a Leverhulme Study Abroad
Studentship between 1998 and 2000. Her thesis is entitled 'Nationalism
and Negotiations of Women's Citizenship'. Her other research interest is
in gender relations in South Asian diasporic communities. She is a
Geography Lecturer at Reading University.

**Katherine Moss** graduated in geography from Cambridge University in 2000.

**Alison Mountz** is a doctoral candidate in the Department of Geography at the University of British Columbia. She researches transnational migration, urban geography and feminist methodology. She has conducted fieldwork in El Salvador, Mexico and the U.S. and is currently studying the reaction of the Canadian state to the arrival of unmarked boats of migrants from the Fujian Province of China to the West Coast of British Columbia.

**Hester Parr** teaches social geography at Dundee University. Her research is focused upon the embodied geographies of people with mental health problems. She is co-editor (with Ruth Butler) of *Mind and Body Spaces: Geographies of Illness, Impairment and Disability* (Routledge). Her work has also been published in the journals *Society and Space, Area, Health and Place* and *Ethics, Place and Environment*. She is currently working on questions of mental health in remote rural communities.

**Samantha Punch** is a Research Fellow in the Department of Applied Social Science at Stirling University. She is currently conducting a post-doctoral study funded by the British Academy on 'Children's Experience of Sibling Relationships and Birth Order in the UK'. Her research interests are in the area of children's geographies and the sociology of childhood and youth. She has previously carried out research on child work in Latin America and Scottish children's problems and coping strategies. Her Ph.D. was an ethnographic study of rural childhoods in Bolivia.

**Sunita Ram** graduated in geography from University College London in 1999 and is taking a professional qualification in personnel while continuing to work in the music industry.

**Pam Shurmer-Smith** is Principal Lecturer in Social Anthropology and Cultural Geography at the University of Portsmouth. She has undertaken fieldwork in Zambia, southern England, Brittany and India. Her current research is on social outcomes of India's liberalizing economy. She is the author of *India: Globalisation and Change* (Arnold, 2000).

**Tracey Skelton** is a Lecturer at Loughborough University. She began her research on Montserrat in 1986 and continues her interest in the island and Montserratian people. She has published in geography and cultural, development and Caribbean studies and is the co-editor of *Culture and Global Change* (Routledge, 1999). Tracey is currently working on a book about the Caribbean to be published by Arnold.

**Susan Smith** is Professor of Geography at the University of Edinburgh. Most of her research is concerned with exposing and exploring geographies of inequality. Her work has contributed to debates on citizenship and social policy, on the problems of racism and gender inequality, on the indignity of victimisation and fear of crime, and on the intractable link between housing and health. She is also interested in how cultural politics work to mediate social exclusion, and in why emotional geographies matter in the pursuit of social well-being.

**Gill Valentine** is a Professor of Geography at the University of Sheffield where she teaches social and cultural geography, and philosophy and methods. Her research interests include: children, youth and parenting; geographies of sexualities; and consumption. She is author of *Social Geographies* (Longman, 2001), co-author of *Consuming Geographies* (Routledge, 1997), and co-editor of *Children's Geographies* (Routledge, 2000), *Cool Places* (Routledge, 1998) and *Mapping Desire* (Routledge, 1995).

# Acknowledgements

We would like to thank all the authors in this volume. All responded to our request for a chapter that focused upon using qualitative methodologies in geography, drawing specifically on their own work, with generosity and enthusiasm. We would also like to thank Peter Jackson for reading and commenting on the Introduction, as well as acknowledging the support and encouragement of our colleagues at UCL and UCN at many stages during the writing and editing of this volume. We are also grateful for the support of all those at Arnold, particularly Laura McKelvie, who originally commissioned the book, and Liz Gooster, who has supported its completion. Finally we would also like to thank our families for their support and encouragement.

We would like to acknowledge the following copyright holders for permission to publish material for which they hold the copyright: Edinburgh University Press, GQ/Conde Nast Publications Ltd.

# 1
# Introduction: doing qualitative research in geography

## Claire Dwyer and Melanie Limb

## Introduction

How do children use space? What are the geographies of people with mental-health problems? How can you research the experiences of new migrants from Mexico in the United States? Or explore competing ideas of identity amongst young Asian women in Britain? How do women and men experience place differently? What are the different discourses and ideas that shape the lives of men and women in the Caribbean? Or in Northern Ireland? Or in rural India? How do people interpret cultural representations of place – through science-fiction novels? Or through popular television shows? What are the 'everyday' geographies underlying socio-economic processes of transnational migration or unemployment, home-ownership or multinational investment? All of these questions are the starting points for pieces of geographical research, many of them described in this book. As geographers our research agendas are provoked by wanting to get behind the 'facts' as they appear to us in everyday life and seek to understand the processes and practices underlying the evidence of change or conflict that we might see around us. For many researchers, gaining a deeper understanding of the complexity of these processes requires a methodology that enables us to engage *in-depth* with the lives and experiences of others. Qualitative methodologies, which explore the feelings, understandings and knowledges of others through interviews, discussions or participant observation, are increasingly used by geographers to explore some of the complexities of everyday life in order to gain a deeper insight into the processes shaping our social worlds.

This book emerged from our discussions about both doing qualitative work in geography and, in particular, supervising undergraduate, masters and Ph.D. students undertaking qualitative research. Both of us are committed to doing qualitative research, Melanie in her work with in-depth discussion groups about environmental meanings (Burgess et al. 1988a, b) and her work on children's geographies (Matthews et al. 1998, Matthews and Limb 1999), Claire in her work on young Muslim women and identity (Dwyer 1999; Laurie et al. 1999) and in her current work on transnational communities and commodity culture. We reflected that as researchers we spend a considerable amount of time thinking about and talking about how the research will be carried out and discussing with others the problems and pitfalls of the work as it progresses. However, this is not a process that is usually reflected in the written outcomes of research, where the space pressure and formal requirements of journal articles tend to relegate methodological issues to a footnote. Thus when you are reading qualitative research in geography you may find little in the *products* that you read that reflect the *processes* of actually undertaking the research – the 'messiness', as Hester Parr puts it, of doing research (Parr 1998). So one of the reasons that we decided to edit this book is to provide an opportunity for students to find out more about these processes of doing qualitative research partly so that you can better understand, evaluate and interpret the geographical work that you read. We also hope that this book will be a useful resource for students, particularly undergraduate students, embarking on a piece of qualitative geographical research for the first time. From our own experiences, we know that students want examples of qualitative research and discussions of qualitative methodologies in geography.

This book is not intended as a step-by-step, 'how to' guide to qualitative methodologies. We expect that readers of this book will have already done some reading in this area and each of the chapters in this book offers guided reading that aims to give more detailed references to the area of qualitative methodology it covers, which we hope will guide readers in some of the specific 'mechanics' of undertaking qualitative methodology. This book is intended to make the process of doing qualitative research more transparent and to open up some of the issues and debates surrounding qualitative methodologies. In each of the chapters the authors draw on their own experiences of using qualitative methodologies in a variety of ways and often across a number of different research projects to reflect on the process of doing qualitative research in geography. What each of the authors seeks to do, in different ways, is to expose some of their own dilemmas and questions in their research decision making to enable other potential researchers to have some insight into the research process. Through these examples we hope to show that qualitative methodologies are a means by which the 'messiness' and complexity of everyday life can be explored by using research methods that do not ignore such complexity but instead engage with it. At the same time we hope this book will demystify the process of *doing* qualitative research.

The book is therefore intended to engage readers, and new researchers, in some of the debates surrounding qualitative methodologies. These debates will, we hope, better inform them in their reading of qualitative research in geography, help them in preparing for their own qualitative research, and open up discussions in lecture theatres and seminar rooms about the processes and issues involved in doing qualitative geographical research.

In this introduction to *Qualitative Methodologies for Geographies: Issues and Debates* we begin with a brief account of qualitative methods in geography. Although not intended to be a detailed history, this section provides a context for the next section, which addresses the question – why choose qualitative methods? Here we outline some of the practical, political and epistemological underpinnings to why and how you might choose to use qualitative methods for your research. The next section of this introduction provides a backdrop to the chapters that follow by outlining some of the issues and debates that are addressed by different authors in the book. Although the book is divided into different subsections, many of these themes cross-cut and so we have chosen to highlight these in the introduction. We also want to stress in this section that the different authors in this volume do not necessarily all agree, and may have different takes on a particular issue. Thus we hope in our introduction to open up these differing views in order to highlight how we want this book to act as a means of debate about qualitative methodologies in geography both within the text itself and beyond. Finally, in this introductory chapter, we offer a brief outline of the book highlighting the different subsections.

## Qualitative methodologies in geography

To begin to write an account of qualitative methodologies in geography is perhaps to anticipate the question – what are qualitative methods? To some extent geography as a discipline has always spanned a wide range of methodological and interpretative approaches. Writing about the antecedents of formal geography Livingstone (1992) stresses the range of different kinds of writings that constituted geographical scholarship, although Domosh (1991) contends that the version of 'scientific' geography that was sanctioned by the nascent Royal Geographical Society in Britain meant the exclusion of some of the more 'qualitative' and 'subjective' accounts detailed by early women travellers. Certainly it is possible to see in the strong emphasis on detailed regional description of the early French School of regional geography (e.g. Vidal de la Blache) or the Welsh school of rural ethnographies (eg. Fleure (see Bowen 1976)) some continuities with more contemporary concerns about qualitative approaches in geography. It was, however, the emergence of humanistic geography in the 1970s as an explicit critique of the prevailing paradigm of geography as a spatial science associated with the Quantitative Revolution that provided the impetus for the exploration of qualitative methodologies by geographers as a means of understanding people's elusive sense of place.

Humanistic geographers sought to challenge the mechanistic and objective approaches that characterized positivism (Entrikin 1976) and emphasized instead the importance of meanings and values held by both researcher and the researched. The pioneers of humanistic geography argued that the intersubjective encounters between researcher and researched should be central to the research process, stressing the need to understand the lifeworlds of individuals and 'the taken-for-granted dimensions of experience, the unquestioned meanings and routinized determinants of behaviour' (Buttimer 1976: 281). Humanistic geographers sought to restore people to the heart of geographical enquiry, arguing that a 'truly human geography' (Tuan 1977) required an understanding of the psychological, emotional and existential attachments that individuals had towards particular spaces, places and landscapes. Humanistic geography drew upon two different kinds of qualitative methodologies. One strand of humanistic geography, characterized in particular by historical and landscape studies, drew upon hermeneutics, focusing in particular on the reading of texts and literature to explore people's associations with and understanding of place (Harris 1978; Meinig 1979). This approach paved the way for the development of a cultural geography concerned with the understanding of cultural landscapes as texts (Daniels 1985; Duncan 1990). The other strand of humanistic geography was more closely allied to social science and drew upon ethnographic methodologies, which were underlain by a phenomenological perspective. Here emphasis was placed on developing 'grounded theories' from careful local studies characterized by participant observation (Ley 1974) and an understanding of the lifeworlds of the researched (Relph 1976; Ley 1977; Seamon 1979). This strand of humanistic geography was particularly concerned with the sub-discipline of social geography, as researchers sought to interpret human experience in its social and spatial settings, drawing on the interpretative methods of ethnography (Jackson and Smith 1984; Smith 1984; Pile 1991). In their landmark text introducing these qualitative methods to geographers, Eyles and Smith (1988: 2) stressed: 'Much of our experience of everyday life is shared and an investigation of this experience and world requires methods which allow the acquisition of "insider knowledge" through interaction, observation, participation in activities and informal interviewing.'

While humanistic geography has undergone a critique (see entry in Johnston et al. 2000), qualitative methodologies have become increasingly important for the practice of human geography. For urban and social geographers, ethnography was recognized as a means by which questions of space and social meaning and social conflict (Ley 1974; Jackson 1985; Keith 1992) could be explored, often bringing geographers into dialogue with anthropologists. For feminist geographers, qualitative methodologies have been particularly important. Although emerging from a Marxist structuralist framework, feminist geography has been centrally concerned with rewriting geography both to incorporate 'the missing half' of a peopled geography and to show that all geographical analysis requires an interpretation of gendered processes and subjectivities (Rose 1993; WGSG 1997; McDowell 1999). Feminist geographers have argued for methodological approaches that are collaborative and non-exploitative and that seek to challenge the unequal power relations between researcher and researched (McDowell 1992), and have thus sought out qualitative methodologies, particularly in-depth interviews, because of the empathetic research encounter that they were felt to engender. However, feminist researchers have also been active in their critique of these possibilities, arguing that ethnographic research methods cannot be assumed to be empowering and indeed raise many significant issues about the vulnerability of research subjects (Stacey 1988). Thus feminist geographers, while still often committed to qualitative methodologies, have been prominent in arguing that the power relations of a research encounter must be recognized and that continuing 'to hanker after some idealised encounter' between researcher and researched is inappropriate (McDowell 1992: 409).

Qualitative methodologies have been significant in the emergence of a range of postmodern and poststructuralist geographies. What these approaches share is a vision of human geography where 'situated' (Haraway 1988) or 'local' knowledges (Geetz 1983) are given prominence over 'grand theory'. Poststructuralist geographies seek to destabilize both the overarching structuralist explanations of Marxist geography and the fiction of the integrated humanist subject that denies multiple, different subjectivities. Current work in human geography encompasses an understanding of the subjectivities of many different others – prompting geographical studies of a wide range of issues, including sexuality (Bell and Valentine 1995), disability and impairment (Gleeson 1998; Butler and Parr 1999), ageing and the life course (Katz and Monk 1993), youth cultures and children (Skelton and Valentine 1998; Holloway and Valentine 2000). Much of this research has been characterized by the employment of qualitative methodologies. An impetus for qualitative research is also evident in an emphasis on post-colonial knowledges in research in non-Western contexts (Miles and Crush 1993; Madge 1994; Radcliffe 1994) where the politics of knowledge construction has been particularly discussed. Again, this is an area of research where geographers have engaged with anthropology particularly in relation to questions about how 'the Other' should be represented (Geetz 1983; Clifford and Marcus 1986; Marcus and Fischer 1986).

The so-called cultural turn in geography (Cook et al. 2000) has been characterized, in part, by the adoption of qualitative methodologies in many different areas of the discipline, such as economic geography (Crang 1994; Lee and Wills 1997; McDowell 1997), population geography (Miles and Crush 1993; White and Jackson 1995), health geography (Dorn and Laws 1994; Kearns 1994), as well as at the interface of human and physical geography, particularly in the study of resource management and public understanding of environmental issues (Harrison and Burgess 1994; Baxter and Eyles 1999). At the same time, the increasing interest in psychoanalytical approaches to understanding the self and others (Rose 1993; Pile 1996; Bondi 1999), and in embodied subjectivities and practices (Pile and Thrift 1995; Longhurst 1997; Nast and Pile 1998), suggests renewed significance for qualitative methodologies within geography.

As this account suggests, qualitative methodologies in geography span a wide range of empirical work and different philosophical and epistemological underpinnings. They also span a variety of different kinds of techniques. First, there are in-depth open-ended interviews, which may be with individuals or groups and which may be conducted once or in a series of meetings – for example, in the construction of a personal life history or a community biography. Next, there are group discussions, which may be single-meeting focus groups or consecutive meetings of in-depth discussion groups. Third, there is participant observation, which may be conducted in a variety of different ways, as full or semi-participant observation, as covert or overt participant observation, as active, even interventionist, or passive participant observation, and which may encompass a range of ethnographic techniques, including the requirement for participants to complete diaries or other interactive exercises. Finally, there are the interpretation and analysis of a wide variety of different kinds of texts including maps, literature, archival materials (such as diaries or letters), landscapes and visual materials including pictures, films, advertisements and dramatic performances. In addition, researchers employing qualitative methods, particularly more ethnographic approaches, will keep personal research diaries in which their own responses to the research process are recorded. Despite this variation in methodology, all of these qualitative approaches share some characteristics – for example, an intersubjective understanding of knowledge, in-depth approach, focus on positionality and power relations, contextual and interpretative understanding. In the next section we outline these characteristics in order to suggest why you might choose qualitative methodologies.

## Why choose qualitative methodologies?

As we suggested above, qualitative methodologies have gained a prominence within human geography, which perhaps suggests that, as students undertaking a research project, you no longer have to defend your choice of qualitative methodologies to sceptical lecturers steeped in a positivist approach to knowledge. What is also increasingly the case is that researchers are deploying a range of different techniques in their research, often combining both quantitative and qualitative approaches (a 'triangulation' of methods), and a number of different studies have emphasized the need for researchers to embrace both quantitative and qualitative measures (Mattingly and Falconer-Al-Hindi 1995; Hubbard 1999). For example, a researcher may choose to use group discussions to explore ideas that individuals have about a particular issue, which then become the basis for the design of a questionnaire for a more extensive study. Or a questionnaire may be used with a large sample group at an initial stage in a research design through which a smaller subsample is then identified for more intensive in-depth interviews. In practice, then, researchers often mix methodologies, reflecting a variety of approaches to tackling a research question (see Valentine, Chapter 3, this volume).

Yet it is also possible to identify particular characteristics of qualitative methodologies that distinguish them from more quantitative methods. Perhaps the most important point is that qualitative methodologies do not start with the assumption that there is a pre-existing world that can be known, or measured, but instead see the social world as something that is dynamic and changing, always being constructed through the intersection of cultural, economic, social and political processes. The emphasis when using qualitative methodologies is to understand lived experience and to reflect on and interpret the understandings and shared meanings of people's everyday social worlds and realities. Qualitative methodologies are characterized by an in-depth, intensive approach rather than an extensive or numerical approach. Thus they seek subjective understanding of social reality rather than statistical description or generalizable predictions. This focus is influenced by the epistemological starting point about the nature of social reality (or the ontological world) held by those using qualitative methodologies; this starting point posits that there is no 'real' world that exists independently of the relationships between researchers and their subjects (Smith 1988). Qualitative methodologies also posit a particular approach to theory building. Rather than being methodologies to test pre-existing theories, qualitative methodologies are used to build 'grounded theory' (Glaser and Strauss 1967) through intensive empirical research so that meanings are clarified and interpreted through the research process.

Thus qualitative methodologies reflect particular understandings of social life and meaning. They also suggest a specific approach to what constitutes the production of knowledge. Researchers who use qualitative methodologies are concerned to recognize and acknowledge the intersubjective values that underlie their research in contrast to the suggestion of 'value-free' or 'dispassionate' methods. Qualitative methodologies are characterized by a relational construction of knowledge between researcher and research subjects and emphasis is placed upon both developing empathy between researcher and researched as well as focusing upon the reflexivity of the research encounter. As Susan Smith makes clear in Chapter 2, a decision to use qualitative methods is also a political choice, since by focusing on the explanations and meanings of the everyday we are placing 'non-dominant' knowledges at the centre of the geographical research agenda. In some cases, a researcher may choose to use a qualitative methodology because she seeks to undertake research that is participatory action research or activist research (see Kobayashi, Chapter 4, this volume). Here the researcher is concerned not simply to empathize with her research subjects but also to help to produce change in their lives. Finally, it is important to recognize that not only are qualitative methodologies characterized by an intersubjective approach to the construction of geographical knowledge in research design and practice, but that this is also an important consideration guiding the writing and dissemination of the research findings, raising issues of positionality and representation, which are discussed in greater detail below.

The choice to use qualitative methods for your research project will be shaped by the dimensions of your research question. However, your choice to use qualitative methods will also depend on how you understand social reality and the philosophical position that you take with regard to the production of knowledge – although this may not be something of which you are aware when you first begin to think about your research question! Your choice of qualitative methodologies will also reflect your own attitude to the subjects of your research study and how you think about the role and responsibility of the researcher. How you frame your research question and what you are seeking to understand in your research will influence which qualitative methods you choose to employ, whether you combine quantitative and qualitative methods, and how you write your research design. These are all steps that are discussed in the first part of this book, particularly in Chapter 3. As you develop your research design and make choices about the kinds of methodologies, you may find yourself confronted with a range of different issues and challenges. In the next section we look at some of the issues and concerns that are debated within the chapters that follow to illustrate some of the questions that using qualitative methodologies provokes.

## Qualitative methodologies: issues and debates

As we suggested above, a philosophical starting point for researchers using qualitative methodologies is that knowledge is situated and partial. This view of knowledge is based upon a recognition of the social world as something that is not fixed or easily known but that is made up of competing social constructions, representations and performances. As Susan Smith argues in Chapter 2, these multiple constructions and representations mean that knowledge is both situated and struggled over. For the qualitative researcher, this recognition of knowledge as partial, situated and socially constructed and contested shapes all stages of the research process. In particular, issues are raised for the researcher about how her own subjectivity and positioning within the research process are acknowledged, how the boundaries between 'insider' and 'outsider' are negotiated and how those with whom the research is undertaken are represented in the written (and other) outcomes of the research.

All of the authors in this volume discuss their own positioning(s) in relation to the research that they undertake. For example, both Stuart Aitken (Chapter 5) and Audrey Kobayashi (Chapter 4) make explicit the ways in which they were motivated to undertake the kind of research that they did because of personal or political concerns. Many of the chapter authors talk about the importance of discussing their own positioning within the research – and about being self-reflective about their role within the research process. As Samantha Punch (Chapter 11) points out, it is important to understand how your research is being shaped by the actions and values of the researcher. Authors reflect on their own positions in different ways: Stuart Aitken in Chapter 5 talks about how he felt it was important to empathize with his interviewees and share his own experiences of parenting with them. In Chapter 12 Hester Parr illustrates how the keeping of her research diary was an important means by which she could record her own feelings and concerns about the research. It is clear that self-reflexivity enters into all stages of the research process: Peter Jackson (Chapter 13) illustrates how he and his co-researchers sought to incorporate their own subject positions into the analytical framework they used to analyse their transcripts. Qualitative researchers are often urged to 'write themselves into the research' and several of the accounts here, particularly David Ley and Alison Mountz's discussion of autobiographical approaches in Chapter 15 and Katy Bennett's suggestion of autoethnography in Chapter 16, suggest how this might be done. Yet these accounts also raise some dilemmas. While all of the authors rightly stress that qualitative research requires an engagement with how the values and subjectivity of the researcher are part of the construction of knowledge, several chapters raise questions about the *limits* of subjectivity. Rose (1997) has argued that calls for reflexivity in qualitative research may be ambitious – that we need to recognize that we cannot be fully aware of, or articulate, our own self-positioning. In this volume both Susan Smith (Chapter 2) and Pam Shurmer-Smith (Chapter 16) caution against the incorporation of 'too much' self-reflection, suggesting that this may make the final written text both exclusionary and self-justified or self-centred. For the qualitative researcher, then, the issue of subjectivity is an important tension that is central to the research process and the production of the written text.

An issue that is highlighted most strongly in the chapters about interpretation and writing, but that all of the authors engage with, is the question of representation. Post-colonial critiques that question the production of knowledge, particularly about marginalized groups, have provoked a 'crisis of representation', inviting considerable reflection both in anthropology (Clifford and Marcus 1986) and in geography (Barnes and Duncan 1992; Crang 1992; Katz 1992; Keith 1992). Given that for many qualitative researchers their aim is to 'recover and centralize marginalized voices', as Hester Parr argues in Chapter 12, such issues of representation are a particular concern. As David Ley and Alison Mountz suggest, two critical questions are raised; first, the role of the researcher involved in what might be seen as an unequal power relation as they interpret the lives of others; and, second, an epistemological question about to what extent *interpretation* is possible and an acknowledgement that all interpretations are 'fictional representations'. Robina Mohammad in Chapter 7 echoes these concerns by asking: 'Who can speak for whom?' For Robina, the authority of the representation rests not on answering the question 'Is this true?' but instead considering 'whose truth' it is. While recognizing that there are many different 'truths', the contributors to this volume emphasize that there are strategies that can be adopted in seeking to represent these multiple meanings. While David Ley and Alison Mountz urge that researchers should remain sceptical of all accounts and produce 'tentative conclusions', Susan Smith suggests that an engagement with the diversity of possible interpretations (and interpretative positions) requires a commitment to listen carefully to conflict, to maximize polyvocality and to discuss subjectivity. Katy Bennett considers some of the different ways in which qualitative researchers might work towards polyvocality, suggesting different literary genres that enable a play of different voices in the written text – even if she acknowledges that geographers do not always find such experimentation easy. While all researchers face the challenges of interpretation in the wake of conflicting ideas and 'clamouring voices', Pam Shurmer-Smith ends her dialogue with Katy with the reminder that, for her, interpretation and writing are not simply the 'representation' of the voices of others but are the construction of new ideas and understandings that go beyond what people say. Again then, what these chapters offer is a dialogue between the authors, both about the challenges faced in seeking to interpret the lives of others and about the responsibilities of the researcher to do just that.

If the writing of oneself into a research practice and the representation of the multiple presentations and representations of others are two issues raised by a recognition of the situatedness of knowledge, a third issue that is important is the negotiation of these positions in the research process itself. Many of the accounts included here show how the authors were involved in negotiations with their informants about their own positioning. This is sometimes articulated as the boundary between 'insider' and 'outsider'; however, many of the authors here illustrate how this opposition is socially constructed and is open to contestation by researcher and researched. Both Tracey Skelton (Chapter 6) and Lorraine Dowler (Chapter 10) highlight some of the ways in which they worked self-consciously to break down their 'outsider' status – and Lorraine emphasizes that this can raise ethical concerns for the researcher. Samantha Punch shows how she occupied a number of different, and sometimes conflicting, subject positions with the research community and she, and her respondents, learned to negotiate her differing roles. Several chapters emphasize that the boundaries between insider and outsider are complex and it is not always easy either to delimit them or to interpret them. Robina Mohammad cautions against any easy binary of 'outsider'/'insider', arguing that how she was positioned by her respondents, and how she chose to position herself, influenced what information she was given and how their 'truths' were represented. This dynamic construction of positioning is illustrated also by Hester Parr, who details how her role with a mental-health user group changed during the course of her research and how she, like Robina, was sometimes confronted about her role by those with whom she worked. While some research accounts may posit straightforward positions of 'insider' or 'outsider' for the researcher, the chapters in this book reveal that the relationship between researcher and researched is likely to be far more dynamic and fluid than this simple binary suggests. Indeed, part of the process of knowledge construction often involves an analysis of how these positionings are constructed and negotiated.

If the question of 'insider'/'outsider' boundary is raised in the doing of the research, Susan Smith raises more fundamental questions in Chapter 12, where she seeks to differentiate between representation/interpretation and performance/presentation. In the chapter itself Susan considers the epistemological differences between these two positions. However, it is also useful to highlight here some of the issues she raises about the practice of doing research. In particular, she urges qualitative researchers either to pay attention, for example, to the practice of conversation, as well as the interview as text, or to learn as much by doing as by talking to people. Several of the chapters in the book, particularly those involving some form of participant observation, emphasize the different insights that such an approach generates. For example, Hester Parr talks about how an office in which she worked became an ethnographic site and the different knowledges that this gave her – although it also required renegotiations of her role – while Samantha Punch describes how her knowledge about children's lives in Bolivia was enriched when she actually participated with them in undertaking their chores.

Finally, in this discussion about the construction and interpretation of knowledge, we want to respond briefly to recent debates about the rigour of analysis in qualitative methods (Baxter and Eyles 1997; see also Bailey et al. 1999). One of the accusations that is sometimes made about qualitative research is that there are dangers that the understandings produced are gained from unrepresentative 'samples' or from carefully selected quotes. As we stated earlier, qualitative research is based upon a 'grounded theory'; as David Ley and Alison Mountz suggest, theory is 'held lightly' and is made accountable to fieldwork. This generation of theory requires both a rigorous approach to analysis as well as an openness to the unexpected, to the challenging of preconceived notions or expectations, as both Hester Parr and Lorraine Dowler demonstrate. We hope that the chapters in this book, particularly those that focus on interpretation, illustrate that qualitative methods require researchers to be both rigorous and *accountable* in their analysis. While Chapters 13 and 14 by Peter Jackson and Mike Crang resist any 'step-by-step' guide to analysis, by opening up the processes that they undertook in their own analysis they not only illustrate how a creative process of theory production is undoubtedly rigorous; they also emphasize that the qualitative researcher needs to be able to show how this process was undertaken. At the same time, the writing of ourselves and others into the research process, as well as detailing the challenges that this entails, is the means by which qualitative research is made accountable.

Having spent some time looking at the issues raised in the chapters in relation to the production of situated knowledge, we now want to turn to some further issues that the chapters raise. One of the concerns that all of the chapters touch upon is how qualitative research is part of a reciprocal relationship between researcher and researched. Both Audrey Kobayashi and Susan Smith stress that qualitative research must be recognized as an intersubjective encounter whereby we and our research become part of people's lives. Audrey's account reminds us that we cannot always know the consequences of our actions – and in her case this results in a painful encounter. Yet this encounter also changes the way she subsequently seeks to do research, seeking an activist approach where her research is used directly by community groups. Other authors trace different reciprocal relationships: Samantha Punch explains how she responds to requests to buy food items when she is in the local town, although she resists other requests. Samantha also ensures that her research is produced in a form that is accessible to her respondents. This is a concern shared by other authors, and David Ley talks about returning with his analysis to share it with those who lived in the community in Philadelphia where he did his research. Returning research to respondents is not straightforward; as Robina Mohammad points out, those she interviewed had conflicting expectations of what they wanted her to represent about their lives. Although returning research to respondents is often advocated by qualitative researchers, Stacey (1988) warns of overburdening respondents with expectations about contributing to your analysis, a view that resonates with Pam Shurmer-Smith's reflections on the responsibility of the researcher. For Stuart Aitken, it is through the process of the interview, which he sees as a collaborative process, that the researcher can begin to challenge the dominant power relations of the research encounter. As Robina Mohammad's experiences indicate, it is important to recognize what the informants may have invested in or expect from the research relationship. While each researcher will decide how the relationship with the research informants, and their expectations, is to be negotiated, it is important that all qualitative researchers approach their research by recognizing that they have responsibilities to those with whom they conduct their research.

These responsibilities include some important issues that are touched upon by several authors but probably need greater emphasis here. These include questions of confidentiality, how access to informants is obtained and how consent from respondents is negotiated by the researcher. In general, a good principle of research practice is that informants are told about how you will use the information that they give you (in terms that make sense to them) and are given the opportunity formally to give, or to withhold, their consent to be part of your research. This process of consent may include their agreement that you will use a pseudonym for them and their home town, for example. In some kinds of research, such as the focus group or prepared interview, it is relatively straightforward to provide respondents with an agreement that they sign, giving you permission to use quotations from their transcripts in your research (indeed, as Gill Valentine points out, this is a legal requirement in the UK). Yet informed consent from respondents is often more complicated. Tracey Skelton highlights the ways in which her interviewees were first introduced to her through a 'gatekeeper'; she points out that in such cases it may be harder for the researcher to establish that the informants have all given their consent to participate. Clearly in some kinds of participant observation, such as that described in Hester Parr's chapter, gaining consent may be difficult and even impossible. In this case, however, Hester's sensitive discussion of these very questions opens up important (and ethical) issues within her research. It is clear that consent is a difficult issue to negotiate and you must make it clear to your interviewees what kind of commitment you can make to them and how you will provide feedback. As a qualitative researcher, it is important that you, in discussion with your advisers, think about issues of consent and the ethics of your research and recognize that it is not always possible to know in advance what impact your research might have.

The importance of maintaining confidentiality with respondents is highlighted in several chapters. Tracey Skelton points out that it is only when her confidentiality is tested and proved that she is able to build relationships with her informants. Similarly, Samantha Punch emphasizes her own responsibilities to maintain the confidence of children, even when this means being evasive in her response to their parents, while Stuart Aitken charts the complexities of maintaining confidentiality when interviewing couples. Stuart's account also reveals a moment when he questions the limits of confidentiality, raising a difficult question about the ethical position of the researcher. While many researchers may not encounter such dilemmas, it is worth thinking about how you will negotiate issues of confidentiality with your respondents.

The final issue we want to raise in this subsection is that almost all of the chapters suggest, in different ways, that qualitative research can often feel confusing and complex. Robina Mohammad uses the metaphor of tangled webs of loose ends that need to be woven into coherence, while Peter Jackson warns of the immense amounts of data that are often generated by qualitative research. It can seem that, in the words of Katy Bennett, you are 'drowning in voices'. At the same time, some of the dilemmas of subjectivity and positionality that we have mentioned above can at times seem overwhelming. It is worth remembering that such feelings are shared by many qualitative researchers! At the same time, much of the richness and depth of qualitative research come from engaging with these contradictions and complexities in seeking to understand new ways of explaining the social world. We hope that the accounts offered here show you how qualitative researchers build such explanations.

## Outline of the book

Although many of the debates and issues that were discussed in the previous section are important across the research process, we have chosen to structure this book around different stages of research design, execution, interpretation and writing up, even though we hope that the previous discussion has emphasized how all of these stages are interrelated and cannot be considered as separate from each other. Thus the main body of the book is divided into six parts: research design; interviewing; group discussions; participant observation and ethnography; interpretative strategies; writing. It is important to state at this juncture that this book does not, therefore, cover all of the many possible qualitative methodologies that have been adopted by geographers. In particular, we do not include chapters that reflect on the interpretation of texts or the use of visual materials. Such methodologies really require a book in themselves, and, for an approach to understanding visual materials, readers are directed to Gillian Rose's book (Rose 2001).

The first part, entitled research design, develops further some of the issues raised in this chapter in the discussion about choosing qualitative methodologies and research design. In the opening chapter, Susan Smith outlines how theory and practice are interlinked and argues that the use of qualitative methodologies is a political choice that can be seen as a strategy that challenges dominant knowledges and structures. Her chapter also sets up a discussion about the distinction between interpreting people's representations of the world and focusing on people's presentations or performances in the world. As she suggests, this distinction may require different kinds of qualitative methodologies and interpretative strategies. Building on this broad discussion, in the next chapter Gill Valentine discusses how to go about choosing the right methodology for a project. Gill covers all the stages of research design, from the construction of research questions to the writing up of the research and its dissemination to readers, who also include the participants in the research. Her chapter illustrates how, in practice, different methodologies are often combined, and Gill illustrates this by drawing on her recent work on children's use of the Internet. The final chapter in this part, by Audrey Kobayashi, presents a very personal account of the use of qualitative methodologies for undertaking what she defines as *activist research*. Audrey raises some of the personal and political dilemmas that she has faced in her research, drawing out in particular issues about positionality and the complex questions that are often raised when we seek to make our research accountable to those with whom we work. While the scope of some of the projects that Audrey discusses may be more ambitious than those encountered by undergraduates, the issues that she raises are crucial for all those who seek to do and understand critical research.

The second part of the book includes three chapters written by researchers using interviews. In Chapter 5 Stuart Aitken reflects on his experiences of interviewing couples as part of a large research project on parenting. Stuart's chapter provides an insight into some of the moral and ethical challenges – as well as interpretative issues – that this research involved. In particular, he discusses some of the ways in which he tried to incorporate his own positioning into the research process, as well as illustrating the different ways in which the participants themselves engaged with the research. This chapter is particularly effective in showing how interviews are a dynamic and intersubjective encounter. In Chapter 6 Tracey Skelton reflects upon experiences of doing research on gender relations in Montserrat. Tracey's chapter raises some of the issues of power and positionality that are encountered particularly acutely in cross-cultural research (whether at 'home' or 'abroad'). Tracey's account describes how she negotiated access to interviewees, considering both the role of gatekeepers and how she worked to gain people's respect and trust. She also discusses the politics of representation, describing some of the dilemmas she faced in publishing her work. The final chapter in this part (Chapter 7) provides an effective contrast to the focus of Tracey's chapter, as Robina Mohammad draws on her experiences of working on a number of different projects involving Pakistani 'communities' in Britain to interrogate the binary of 'insider'/'outsider' in the construction of a researcher's positionality in the field. Robina emphasizes the different and shifting positions of both herself and her research subjects within a negotiated process of knowledge production. Robina's account highlights both the limits to 'reflexivity' and the multiple 'truths' that are produced through the research encounter, while arguing that the choice of which 'truth' is represented remains the responsibility of the researcher.

The third part of the book looks at the qualitative methodology of group discussions. In Chapter 8 Tracey Bedford and Jacquelin Burgess outline the practice of focus groups, drawing on research commissioned for the British Retail Consortium. Tracey and Jacquelin use their case study example to outline all the stages of working with focus groups, from the recruitment of groups, the conducting of the group discussions, to how to go about analysing the transcripts that these group discussions produce. In Chapter 9 James Kneale provides an insight into the use of in-depth groups through a discussion of his research on science-fiction readers. James outlines some of the challenges he faced in the setting-up and running of groups and provides a particularly insightful analysis of how he dealt with conflicts within the discussion groups. He uses this discussion to illustrate how the dialogical production of meanings within groups can be interpreted and analysed.

In the fourth part we focus on the qualitative methodologies of participant observation and ethnography. In the opening chapter of this part (Chapter 10), Lorraine Dowler describes her ethnographic work in Belfast. Her account, which highlights in particular how respondents were accessed and how she herself was integrated into the research context, outlines some of the advantages that participant observation offers for understanding complex situations. Lorraine discusses some of the ways in which her own assumptions about 'terrorists' were changed through the ethnographic research. The next chapter in this part, by Samantha Punch, draws on her research with children in rural Bolivia. Samantha emphasizes that understanding the social worlds of children, particularly within a very different cultural milieu from her own, required the creative employment of a range of different qualitative methods. She shows how a variety of participatory techniques with children were used to supplement the understanding she gained from informal interviews and semi-participant observation. Samantha's account is not only a rich example of the benefits of combining research methods but also illustrates that it is *by doing* (for example, collecting water with the children) that she was better able to understand the experiences of those to whom she talked. In the final chapter of this part (Chapter 12) Hester Parr describes the different strategies she used in researching people with mental-health problems through ethnography. She draws upon two different 'ethnographic encounters', which vary in the extent to which they are overt/covert, one in which her 'enthnographic site' emerged, the second a discussion of 'unsuccessful' ethnography. Through her detailed analysis of these two ethnographic encounters, Hester illustrates how academic knowledge is produced in ways that are often unexpected for the researcher, demonstrating the need for a reflective and reflexive approach to understanding and interpretation.

The fifth part of the book contains three chapters that focus, in different ways, on the question of interpretation. In Chapter 13 Peter Jackson reflects on how he and his colleagues made sense of the qualitative data (both group and individual interview transcripts) generated in a recent project on men's 'lifestyle' magazines. Peter describes the processes of coding and the creation of theoretical themes undertaken by himself and others within the research team. By taking us through the different steps in interpretation and understanding, as well as describing some of the ultimately unproductive avenues that he explored, his account illustrates the iterative processes of qualitative analysis. In Chapter 14 Mike Crang opens up the interpretation process he underwent in the analysis of group transcripts from local history groups in Bristol. Mike's account provides an insightful contrast to Peter's chapter, not only in terms of the processes involved, particularly his use of a computer program, but also in outlining a rather different sequence of interpretation. Both chapters provide readers with examples of the 'mind maps' and thematic codes that their analysis produced. Together these chapters provide an illustration of the rigorous approach to interpretation that must be undertaken with qualitative data while also suggesting that there are no hard-and-fast rules to such analysis. The last chapter in this part, by David Ley and Alison

Mountz, offers a broader meditation on interpretation, beginning by outlining both the ethical and epistemological questions that have been provoked by the 'crisis of representation'. The authors then draw on their own research projects – in David's case discussing research conducted in 1970s Philadelphia and in 1990s Vancouver, in Alison's focusing on recent work on Mexican migrants in the United States – to consider how their own interpretations were constructed and, particularly interestingly, subjected to scrutiny. Finally, they reflect on efforts by ethnographers to insert themselves into the stories that they tell, emphasizing the personal as well as the political dimensions of interpretation.

The last of the main parts of this book deals with writing. In fact many of the previous chapters have also discussed issues of writing or representation and this is important, as we would want to stress that writing is not something that occurs only when all the data are collected but is an ongoing part of the process (for example, through the research diary). Indeed, although taking writing as their starting point, these last two chapters are also inextricably about interpretation, illustrating that these processes are interlinked in the creation of knowledge and the representation of others. In Chapter 16 Katy Bennett and Pam Shurmer-Smith discuss the tensions of representing others and the contradiction between a seemingly fixed text and the 'clamouring voices' of the researched. Their 'writing conversation' seeks to mirror these dilemmas, as they reflect on the role of subjectivity in academic writing and the responsibility of the researcher. Ruth Butler's chapter also discusses the role of the personal in academic writing, drawing on her own experiences 'as a woman with a visual impairment' writing about disability, and raises questions about some of the ways in which the personal is used in academic writing.

The book ends with a final part containing five vignettes written by undergraduates, which reflect on the research methods they used for their undergraduate dissertations. In the first vignette, Katharine Moss uses a combined methodology of questionnaires and in-depth interviews to explore girls' aspirations and their schooling experiences in South India. Then, through in-depth interviews and participant observation, Sunita Ram explores the role of black music in the construction and contestation of hybrid Asian identities. Third, Annabelle Aish uses group discussions and interviews to investigate the experiences of househusbands and the ways in which they construct their masculine identities. Sarah Corke also uses group discussions to analyse audience responses to the television programme *Goodness Gracious Me*. Finally, Philippa Capps discusses her use of participant observation to explore the geographies of a retail workplace. These accounts suggest some of the diversity of projects that undergraduates might undertake that adopt qualitative methodologies. The authors draw out both the issues involved in their choice of methods as well as the ways in which they themselves were reflexively constituted within the research.

Although they are different from each other, the perspective that unites all the authors is that they draw upon their own research experiences very directly in writing about doing qualitative research. By way of conclusion we want both to acknowledge and to applaud the generosity of the contributing authors to this collection for responding to our request to write about their research experience in such honest, reflective and often highly moving accounts. We hope that readers of this book will share our enthusiasm for the insights that these accounts produce and that they will also motivate you in your own research.

## Key references

Cloke, P., Philo, C. and Sadler, D. 1991: *Approaching human geography*. London: Paul Chapman.

Eyles, J. and Smith, D. (eds) 1988: *Qualitative methods in human geography*. Cambridge: Polity Press.

Johnston, R.J., Gregory, D., Pratt, G. and Watts, M. 2000: *The dictionary of human geography*. Oxford: Basil Blackwell.

## References

Bailey, C., White, C. and Pain, R. 1999: Evaluating qualitative research: dealing with the tension between 'science' and 'creativity'. *Area* 31(2), 169–78.

Barnes, T. and Duncan, J. 1992: *Writing worlds*. London: Routledge.

Baxter, J. and Eyles, J. 1997: Evaluating qualitative research in social geography: establishing 'rigour' in interview analysis, *Transactions – Institute of British Geographers* NS 22(4), 505–25.

———— 1999: The utility of in-depth interviews for studying the meaning of environmental risk. *Professional Geographer* 51(2), 307–20.

Bell, D. and Valentine, G. 1995: *Mapping desire*. London: Routledge.

Bondi, L. 1999: Stages on journeys: some remarks about human geography and psychotherapeutic practice. *Professional Geographer* 51(1), 11–24.

Bowen, E. G. 1976: Herbert John Fleure and western European geography in the twentieth century. Reprinted in Bowen, E. G., *Geography, culture and habitat: selected essays (1925–1975)*. Llandysul: Gomer Press.

Burgess, J., Limb, M. and Harrison, C. M. 1988a: Exploring environmental values through the medium of small groups: 1. Theory and practice. *Environment and Planning A* 20, 309–26.

———— 1988b: Exploring environmental values through the medium of small groups: 2. Illustrations of a group at work. *Environment and Planning A* 20, 457–76.

Butler, R. and Parr, H. (eds) 1999: *Mind and body spaces: geographies of disability, illness and impairment*. London: Routledge.

Buttimer, A. 1976: Grasping the dynamism of the lifeworld. *Annals of the Association of American Geographers* 66, 277–92.

Clifford, J. and Marcus, G. 1986: *Writing culture: the poetics and politics of ethnography*. Berkeley and Los Angeles: University of California Press.

Cook, I., Crouch, D., Naylor, S. and Ryan, J. 2000: *Cultural turns/geographical turns: perspectives on cultural geography.* Harlow: Prentice Hall.

Crang, P. 1992: The politics of polyphony: reconfigurations in geographical authority. *Environment and Planning D: Society and Space* 10(5), 527–49.

—— 1994: It's showtime: on the workplace geographies of display in a restaurant in southeast England. *Environment and Planning D: Society and Space* 12, 675–704.

Daniels, S. 1985: Arguments for a humanistic geography. In Johnston, R. (ed.), *The Future of Geography.* London: Methuen, 143–58.

Domosh, M. 1991: Towards a feminist historiography of geography. *Transactions of the Institute of British Geographers* NS 16, 95–104.

Dorn, M. and Laws, G. 1994: Social theory, body politics and medical geography. *Professional Geographer* 46, 106–10.

Duncan, J. 1990: *The city as text: the politics of landscape interpretation in the Kandyan kingdom.* Cambridge: Cambridge University Press.

Dwyer, C. 1999: Veiled meanings: young British Muslim women and the negotiations of differences. *Gender, Place and Culture* 6(1), 5–26.

Entrikin, J. N. 1976: Contemporary humanism in geography. *Annals of the Association of American Geographers* 66, 615–32.

Eyles, J. and Smith, D. (eds) 1988: *Qualitative methods in human geography.* Cambridge: Polity Press.

Geetz, C. 1983: *Local knowledge: further essays on interpretive anthropology.* New York: Basic Books.

Glaser, B. and Strauss, A. 1967: *The discovery of grounded theory: strategies for qualitative research.* Chicago: Aldine.

Gleeson, B. 1998: *Geographies of disability.* London: Routledge.

Harasway, D. 1988: Situated knowledges: the science question in feminism and the privilege of partial perspective. *Feminist Studies* 14(3), 575–99.

Harris, R. C.1978: The historical mind and the practice of geography. In Ley, D. and Samuels, M. (eds), *Humanistic geography: prospects and problems.* London: Croom Helm, 123–37.

Harrison, C. M. and Burgess, J. 1994: Social constructions of natures: a case study of the conflicts over Rainham Marshes SSSI. *Transactions of the Institute of British Geographers,* 19(3), 291–310.

Holloway, S. L. and Valentine, G. 2000: *Children's geographies: playing, living, learning.* London: Routledge.

Hubbard, P. 1999: Researching female sex work: reflections on geographical exclusion, critical methodologies and 'useful' knowledge. *Area* 31(3), 229–38.

Jackson, P. 1985: Urban ethnography. *Progress in Human Geography* 9, 157–76.

—— and Smith, S. J. 1984: *Exploring social geography.* London: Allen Unwin.

Johnston, R. J., Gregory, D., Pratt, G. and Watts, M. 2000: *The dictionary of human geography.* Oxford: Basil Blackwell.

Katz, C. 1992: All the world is staged: intellectuals and the project of ethnography. *Environment and Planning D: Society and Space* 10, 495–510.

—— and Monk, J. (eds) 1993: *Full circles: geographies of women over the life course.* New York: Routledge.

Kearns, G. 1994: Putting health into place: an invitation accepted and declined. *Professional Geographer* 46, 111–15.

Keith, M. 1992: Angry writing: (re)presenting the unethical world of the ethnographer. *Environment and Planning D: Society and Space* 10, 551–68.

Laurie, N., Dwyer, C., Holloway, S. and Smith, F. 1999: *Geographies of new femininities.* London: Longman.

Lee, R. and Wills, J. 1997: *Geographies of economies.* London: Arnold.

Ley, D. 1974: *The black inner city as frontier outpost: image and behaviour of a Philadelphia neighbourhood.* Monograph Series 7. Washington DC: Association of American Geographers.

—— 1977: Social geography and the taken-for-granted world. *Transactions of the Institute of British Geographers* NS 2, 498–512.

Livingstone, D. 1992: *The geographical tradition.* Oxford: Basil Blackwell.

Longhurst, R. 1997: (Dis)embodied geographies. *Progress in Human Geography* 21, 486–501.

McDowell, L. 1992: Doing gender: feminism, feminists and research methods in human geography. *Transactions of the Institute of British Geographers* NS 17(4), 399–416.

—— 1997: *Capital culture.* Oxford: Basil Blackwell.

—— 1999: *Gender, identity and place.* Cambridge: Polity.

Madge, C. 1994: The ethics of research in the 'Third World'. In Robson, E. and Willis, K. (eds). *Postgraduate fieldwork in developing areas.* DARG Monograph No. 8. London: Royal Geographical Society/Institute of British Geographers, 91–102.

Marcus, G. and Fischer, M. M. J. 1986: *Anthropology as cultural critique: an experimental moment in the Human Sciences.* Chicago: University of Chicago Press.

Matthews, H. and Limb, M. 1999: Defining an agenda for the geography of children: review and prospect. *Progress in Human Geography* 23 (1), 61–90.

—— —— and Taylor, M. 1998: The geography of children: some ethical and methodological considerations for project and dissertation work. *Journal of Geography in Higher Education* 22 (3), 311–24.

Mattingly, D. and Falconer-Al-Hindi, K. 1995: Should women count? A context for the debate. *Professional Geographer* 47 (4), 427–37.

Meinig, D. 1979: *The interpretation of ordinary landscapes*. New York: Oxford University Press.

Miles, M. and Crush, J. 1993: Personal narratives as interactive texts: collecting and interpreting migrant life-histories. *Professional Geographer* 45, 83–95.

Nast, H. and Pile, S. 1998: *Places through the body*. London: Routledge.

—— et al. 1994: Women in the field: critical feminist methodologies and theoretical perspectives. *Professional Geographer* 46, 1, special issue.

Parr, H. 1998: Mental health, the body and ethnography. *Area* 30 (1), 28–37.

Pile, S. 1991: Practising interpretive geography. *Transactions of the Institute of British Geographers* 16, 458–69.

—— 1996: *The body and the city: psychoanalysis, space and subjectivity*. London: Routledge.

—— and Thrift, N. (eds) 1995: *Mapping the subject: geographies of cultural transformation*. London: Routledge.

Radcliffe, S. 1994: (Representing) post-colonial women: authority, difference and feminisms. *Area* 26 (1), 25–32.

Relph, E. 1976: *Place and placelessness*. London: Pion.

Rose, G. 1993: *Feminism and geography*. Cambridge: Polity.

—— 1997: Situated knowledges: positionality, reflexivities and other tactics. *Progress in Human Geography* 21 (3), 305–20.

—— 2001: *Visual methodologies*. London: Sage.

Seale, C. 1998: *Representing society and culture*. London: Sage.

Seamon, D. 1979: *A geography of the lifeworld*. London: Croom Helm.

Skelton, T. and Valentine, G. (eds) 1998: *Cool places: geographies of youth cultures*. London: Routledge.

Smith, S. J. 1984: Practising humanistic geography. *Annals of the Association of American Geographers* 74, 353–74.

—— 1988: Constructing local knowledge: the analysis of self in everyday life. In Eyles, J. and Smith, D. (eds), *Qualitative methods in human geography*. Cambridge: Polity, 17–38.

Stacey, J. 1988: Can there be a feminist ethnography? *Women's Studies International Forum* 11 (1), 21–7.

Tuan, Y.-F. 1977: *Space and place: the perspective of experience*. London: Edward Arnold.

WGSG (Women and Geography Study Group) 1984:, *Geography and gender: an introduction to feminist geography*. London: Hutchinson and Explorations in Feminism Collective.

—— 1997: *Feminist geographies: explorations in diversity and difference*. London: Longman.

White, P. and Jackson, P. 1995: (Re)theorising population geography. *International Journal of Population Geography* 1 (2), 111–23.

# PART I

## Research design

# 2
# Doing qualitative research: from interpretation to action
## Susan Smith

## Introduction

This book is about the diverse possibilities for practising qualitative methodologies in human geography. That there *are* so many options is testimony to the vitality of qualitative social science. This means that to write 'I am doing qualitative, rather than quantitative, research' is no longer sufficient to convey the flavour of a work. We now need to know what kind of qualitative study is involved, and, crucially, why it was chosen. Of course, pursuing any kind of qualitative enquiry is still about adopting a particular kind of position in relation to studied subjects. I shall discuss that positioning first, arguing that choosing qualitative methods is not so much a philosophical or theoretical conundrum as a self-conscious political act: a statement about how you believe the world is and should be.

In the second section of the chapter I want to go on to consider which qualitative methodologies might best be used, when; and to show *how* they can be used to achieve particular research aims. This does not include a comprehensive tour of the options – there is plenty of space later in the book to see precisely what is on offer. Rather I want to make one main argument. My suggestion is that a key consideration for researchers today is the extent to which (for the purposes of any particular study, or part of a study) they are: (1) interested in how people see, experience and make particular representations of the world as it is (and has been); and (2) interested in how people 'do' things, in how they (which includes we) make the world as it goes along, as it becomes. In one sense, of course, these are two sides of a single coin. But methodologically they do imply, or at least crystallize around, two rather different sets of research strategies.

In the 1990s there was a frightening array of philosophical, conceptual and theoretical terms embedded in the qualitative research literature. Even the same empirical study might be discussed by different researchers in very different ways using quite distinct vocabularies. An ethnography, for example, may be read as a text or lived as a performance; a sentence can be classified according to content or negotiated dialogically. Therefore, in reducing this whole chapter to two 'big' points (namely, first, that choosing qualitative methods is a political rather than a philosophical issue and, second, that

**Fig. 2.1** *Mr and Mrs Andrews, Thomas Gainsborough*
*Reproduced with permission from The National Gallery*

qualitative methods cluster according to whether they are concerned with eliciting/interpreting representations, on the one hand, or with documenting and doing presentations, or performances, on the other (see Fig. 2.1.) I am guilty of making something that is complex, messy, difficult and demanding seem too straightforward, too neat. I am also guilty of making you think about qualitative methods in the same way that I do, when in fact this is an area that is very personal, highly contested and rarely 'right' or 'wrong'. Yet it is partly because the field can be so intimidating that I am trying to offer a relatively straightforward starting point; and it is precisely because the once-neat set of links between philosophy, theory and method is now being unpicked that I am rejecting the notion of a comprehensive overview of 'traditional' ontological and epistemological concerns. Because this is very much a practical text, I have also tried to draw examples from my own research as much as possible. This means that I shall offer a rather partial account of a wide-ranging literature. But some partiality is inevitable for this is not an enterprise that will ever be finished, definitive or closed to other experiences and voices.

## The politics of qualitative research

It is usual to suggest that qualitative methods are a useful way of proceeding when we are interested in a multiplicity of meanings, representations and practices. That is, qualitative methods become appropriate when we have abandoned those foundationalist ideas that suggest there is a real world, a big picture, an ultimate truth; when we have given up on the idea that totalizing accounts of how things are can work; when we recognize that such accounts are not only impossible but unethical. So it might be said that qualitative methods (certainly of the type discussed in this volume) stem from a particular if rather broadly cast belief (or ontological position): that the world is not real in a fixed, stable or predictable way; that it is not entirely accessible; and that it does not appear empirically the same to everyone, no matter how carefully we look. Rather, qualitative methods presume the world to be an assemblage of competing social constructions, representations and performances. These assemblages – these various ideas about how the world is and should be, these various practices that move us from the present to the future – are of interest to human geographers for two reasons. First, they take place. Second, the ideas that stick and the practices that work are not fixed essential features in the landscape of a 'real world' . Rather they testify to the power relations, struggles and negotiations that allow particular versions and visions of the world to be realized in particular places at particular times.

It follows that no one – neither researcher nor researched – can fully know the world, or fully be detached from the construction of knowledge. Knowledge is situated and struggled over, so the broad ontological position of qualitative research comes with a particular epistemological mandate – a particular set of ideas about just what can be known and about how this knowledge can be accessed or made. This mandate is one that requires researchers to recognize the extent to which they are immersed in, rather than detached from, the production of knowledge. For the methods discussed in this book, this means that researchers can expect to engage in a close relationship with the people whose lives they encounter.

This all helps us to understand some of the essential differences between qualitative and quantitative approaches; between positivistic, realist, scientific models of enquiry, on the one hand, and the humanistic, interpretative traditions, on the other. It does not help much in making the decision between ethnographic or interview-based methods; between case study or focus-group research, and so on. But it does, nevertheless, imply that using any of these is about having made an important set of choices. And my argument in this section is that these choices are not so much about philosophical and methodological abstractions as about political and ethical practicalities. Let me elaborate.

All qualitative methods recognize the relevance and importance of 'lay' or 'folk' perspectives on the practicalities of everyday life. We choose these methods, then, as a way of challenging the way the world is structured, the way that knowledges are made, from the top down. We are therefore adopting a strategy that recognizes the diversity of human experience, that addresses the complexity of how lives are lived, and that confronts the fact that people's characteristics and experiences do not group into neat mappable parcels or tidy policy-relevant units. We are, then, adopting *a strategy that aims to place non-dominant, neglected, knowledges at the heart of the research agenda.* Whether we always succeed in doing this is a different question. But, whether tacit or self-conscious, this political goal of unsettling the status quo, of redefining what relevant, useful, legitimate knowledge is, is what these research strategies are about. It is hardly surprising then that such research has been closely associated with the development of feminist, anti-racist and post-colonial knowledges, and indeed with the production of knowledges rooted in the experiences of a wide range of socially excluded or marginalized groups.

Let me take as one example my work with Helen Roberts, Carol Bryce and Michelle Lloyd on child accidents (Roberts et al. 1993, 1995). This took place at a time when the British government had child accident reduction as one of its major targets in health policy, and when the primary vehicle to this end was health education. If parents knew more about what puts their children at risk, the reasoning went, they would take better preventative measures and the accident rate would fall. Our research happened, however, not because of anything the government said, but because a group of mothers in a Glasgow council estate contacted Glasgow university to ask for help. Their concern was how to tackle what they believed to be a high child accident rate in their own community. This was the first step in a study that was to show that parents have a somewhat better knowledge of risks, and of what to do about them, than the health education model implies.

As we worked through a series of group interviews and a set of qualitative accounts of accidents and 'near' accidents (together with an epidemiological study to specify just what the accident rate was), the root of the problem emerged as a hazardous built environment, together with parents' inability to unlock the (quite modest) scale of resources required to make at least some parts of this environment safe. The relatively high rate of child accidents in the locality turned out not to be a product of the limited knowledge base, much less the routine neglect, of parents. On the contrary, precisely because they live in such dangerous spaces, parents in high-risk areas know a lot about what causes child accidents. Ironically, they know a lot too about accident prevention – because they do this every day. Indeed, it was our opinion that, without this local 'lay' knowledge, the local accident rate would have been much higher. The miracle in such dangerous spaces is that most children remain safe, and escape injury, for most of the time. This miracle happens day after day simply because parents are so constantly alert to danger and so consistently aware of risk. In short, the results of this research suggest that, rather than viewing parents as the source of the problem, policy-makers would do better to regard them as part of the solution. These neglected knowledges are the ones that count – the ones that, if acted on, could really make a difference to child safety and well-being.

The qualitative methods at the heart of this book are not only about allowing 'other' voices to speak. They are also strategies of encounter – about how we and our research become part of other people's lives. And this poses both an ethical dilemma and a political opportunity. The ethical dilemma is what currently attracts most concern. This is for good reasons, because what we are recognizing here is that researchers have a role in making other people's knowledges; the analyst becomes 'a co-constructor of stories about a world open to different interpretations according to vantage point' (Dyck 1999: 245). Reflexivity and critical self-reflection are, rightly, an important part of the research process, and some of the ideas developed by Baxter and Eyles (1997) and by Bailey et *al.* (1999), on the practice of reflexive project management, on the value of providing a methodological paper trace for key research decisions, and on the full documentation of how research practices are enacted, urgently need developing and implementing. Nevertheless, too much agonizing about this can produce a kind of textual narcissism in academic writing, where nothing can be said without a string of qualifications, provisos and auto-critique. Here I would agree with Thrift (1996) that the trend towards texts accompanying texts can be patronizing to the reader. More important still, this trend towards over-self-justification can also be a means by which authors claim more not less authority, so inadvertently undermining the credibility of what the non-academics have to say.

On the ethics of this research, there is a fine line to tread. My main concern here, though, is that a preoccupation with the ethical dilemma often comes at the expense of (rather than in conjunction with) a thoroughgoing engagement with the political opportunity this kind of research affords (indeed demands). Thrift alludes to this when he talks about 'the one special responsibility that I do think academics have, which is to multiply the communicative resources that people have available to them' (1996: p. xi). Offe (1984) makes a similar point when he argues that critical social science is about framing research and raising questions in ways that are accessible to *all* those who might make a difference. But, given that doing qualitative research at all implies a political choice, it seems to me that, however small the 'p' in political, there is more to it than this. Keith (1992) and Back and Solomos (1993) argue this point in relation to the emotive politics of race. Yet, for human geography more generally, even though we know that doing qualitative research *is* about changing the world, we still harbour that nagging suspicion that these changes should be minimized. My question, then, is this: if researchers should not be changing the world, what is the point of engaging with it at all? Qualitative research is a mode of interference. If this interference is 'wrong', such approaches should be discontinued; they must be unethical. If it is 'right', a lot more documentation and debate are needed on where this interference is going, whose interests it advances, what form it takes and why it is important.

## Doing qualitative research

Choosing qualitative research is about engaging with a particular kind of politics. But qualitative research refers to a multitude of practices. We may, in this book, be concentrating only on strategies that involve some kind of direct encounter between researchers and those they are studying, we may 'only' be studying social relations as mediated through everyday experience in everyday space, but this still offers a bafflingly wide range of possibilities. They span all kinds of group and individual interviews (which may be organized around a talk-based interview guide, around visual images, musical exerpts, diaries, life grids and so on), and various ethnographic methods (from participant observation to action research, from 'thick description' to performance events).

Many factors determine which of these various qualitative approaches is adopted. Cresswell (1998) discusses some of them in his guide to choosing among five key qualitative research traditions. His basic suggestion is that which method you choose depends on what you want to know or do: select a biography to study a person, a grounded theory to develop a theory, an ethnography to study behaviour in a group, a phenomenology to examine a phenomenon and so on. There are other classifications and typologies of qualitative research that are pragmatic in this kind of way. And to some extent pragmatic considerations must dominate methodological decisions. This kind of research is, after all, about the intrusion of academic careers into other people's lives. Adopting a strategy that seems theoretically well-founded but that irritates, offends or upsets can rarely be justified. Likewise, approaching people with ideas that are in vogue with the prevailing philosophical (or anti-philosophical) thinking, but that have no appeal to participants, is never going to work.

Being concerned with these practical problems is clearly important, as the rest of this book will show. In the remainder of *this* chapter, however, I want to explore one key *theoretical* consideration that might inform or inspire particular qualitative methodological choices, and that might suggest particular ways of doing qualitative research. This key consideration is about the difference between representations/ interpretations, on the one hand, and 'doings'/presentations/performances, on the other. I will take them in turn.

### Interpretations: writing and reading the landscape

Perhaps the majority of qualitative research published up to 2000 implicitly or explicitly regards social life as an ensemble of texts to be read and interpreted. Once Ricoeur (1971) had argued that all types of cultural production – writing, landscapes, maps, festival events and so on – have textlike qualities, the scene was set for a 'textual' and linguistic turn in social research. After Barthes (1977) had announced the metaphorical 'death' of the author, it was clear that, once 'written', texts ceased to convey just, or even, the meanings the author intended. Rather, they became open to contestation, to negotiation, to multiple readings.

The classic accounts of reading cultures as texts are generally attributed to Clifford Geertz, an anthropologist steeped in the ethnographic tradition. This textual turn in anthropology is discussed by Howes (1991). There *is* a well-developed ethnographic tradition in geography, and much of what I am arguing in this section applies to most of it. But it is probably fair to say that various interview-based approaches have been the more prominent strategies of encounter in yielding key texts for interpretation within human geography. (It is important to note that the textual turn is not limited to strategies of encounter. Some fascinating work by historical geographers is based on other kinds of 'readings' and Duncan's interpretation (1990) of the texts of the Kandyan landscape is a seminal work in the textual tradition.) So, for work within the scope of the present volume, the qualitative interview transcript is probably the single largest source of qualitative data currently held by geographers. Qualitative interviews are methodologically appealing because they allow a wide range of experiences to be documented, voices to be heard, representations to be made and interpretations to be extracted. Open-ended, qualitative interviews are, after all, the obvious way of allowing people to speak for themselves about their own views and experiences of the world. Such interviews, whether conducted with groups or with individuals, are moreover widely regarded as a way of accessing those knowledges neglected in an earlier period of more 'authoritative' social research.

In keeping with this democratizing spirit, much of the literature on qualitative interviewing is concerned to ensure that the design and conduct of such interviews are empowering for participants (Bowes 1996; Morgan 1996). Focus-group research, for example, which has been enthusiastically adopted by the market-research industry, can easily be reduced to a closed format that has the same positivistic trappings as the formal questionnaire survey (Johnson 1996). On the other hand, focus groups, as well as more extended group and individual interviews, can be designed to empower participants and democratize the research process in a much more radical way, allowing participants not only to speak for themselves, but also to have a say in interpreting the 'data' they produce (Burgess et al. 1988; Roberts et al. 1995).

Once this more radical way is adopted, and once a host of issues surrounding plausibility, validity, rigour and so on are addressed (see Smith 1981; Baxter and Eyles 1997; Bailey et al. 1999), the key issue still remains: what does this textual, interpretative approach have to offer? The short answer is this: behind perhaps the majority of interview-based qualitative research, there is an idea that what the research is for is to provide an interpretation of subjects' intepretations of theirs and others' representations of what the world is like. What we are doing here is accessing the world as people think it is and has been. We are accessing a representation (a vision, an image, an experience) of a text (the world of lived experience) through a text (the interview transcript) that is itself open to interpretation.

Now, although it is in practice highly complicated and fraught with caveats, this textual approach has some important merits. Consider, for example, a project I am working on with my colleagues Liz Bondi, Hazel Christie, Shonagh MacEwen and Moira Munro on house-price inflation in Edinburgh. The majority of the existing literature on house-price instability is produced by housing economists. It is the economics of supply and demand – mediated by interest rates, demographic change and wages – that explains, statistically, the cycles of housing booms, and slumps, in a country like the UK. What we find missing in that literature, however, is any real sense that geography matters, and any ability to predict or account for just when and where house prices become unstable. The reason for this is that no one has asked the people most directly involved in buying and selling houses *why* and *how* they pay what they do for particular kinds of property in particular kinds of area. Our project, on the other hand, is designed to do just this – to flesh out the anatomy of a housing boom as it is constructed, represented and experienced by both households and the professionals they rely on. We think that qualitative interviews with these various actors will provide insight not only into how housing booms start and what drives them on, but also into how economic processes are embedded within people's much wider social and cultural lives.

I think this project is important, then, because it acknowledges that housing booms are not impersonal events that can be explained in terms of macroeconomic factors interacting with the financial institutions that make up 'the' housing market. This economic vision is just one set of representations of what housing markets are about. Our research suggests that housing markets are also about people's lives, their hopes, their aspirations, their despair. They are about a richly textured world that people who buy and sell their homes represent to one another, interpret for one another, and act in (and change) on the basis of these constructions and representations.

None of this would matter much if representations were simply a set of benign images detached from the material world. However, representations are implicated in the constitution of the world. How the world is, is partly about the struggle to make a particular set of ideas about how that world works stick; it is about whose interests are built into the dominant vision of things. The economic representation of house price instability is currently dominant – it is what drives housing policy and institutional practice. The whole point of our project is to allow other imagined geographies to be spoken about and realized. The idea is to replace a single dominant representation with a variety of other interpretations, with competing representations, with a plurality of views, which might, once articulated, begin to challenge the status quo; which might one day play a part in changing how, and for whom, the housing market works.

Now, working with text and language as ways of representing the world, and of making it like it is, has been an important challenge for qualitative research, not least because it has required us to tackle the valid anxiety that 'we' should not be speaking for 'them'; that 'they' should represent 'themselves' (Katz 1992). This concern, born of post-colonial and feminist critiques of knowledge, indeed posed a major crisis of representation – of who should represent whom and how – in social research. Analysts became interested not just in what a landscape, a scene, an event represents, but in how it represents different things to different people, and also why some people's points of view and experiences are not represented at all. Hence the transition from single authorial interpretations to dialogism and polyphony, from one person interpreting a text to lots of people talking about it, and creating ever more texts and representations.

However, this tradition does have its limits. As Silverman (2000) recognizes, the fashionable identification of qualitative method with an analysis of how people 'see things' ignores the importance of how people 'do things'. The anthropologist David Howes (1991) makes a related point: that examining social and cultural life as an assemblage of texts is, literally, a 'non'-sense: a question of reading and writing cultures rather than one of sensing and experiencing them. There is some concern, then, that, in the conduct of qualitative research, text, language and vision – the stuff of representation

– have dominated practice, talk and a rounded sensory experience. Indeed, we might say that there is now a second crisis of representation – a concern about the limits of representability, a question mark over the fact of representing at all. As Howes (1991: 69), puts it: 'Events or social activities do not possess the same stability as texts. This raises the question of whether the hermeneutic method [a body of rules for the interpretation of written documents] can or even should have been extended from the study of texts to the study of action.'

## Presentations: ways of being and doing

That qualitative research should seem deficient in its attention to practice is ironic, because such research is above all about intimate encounters. It is the only collection of methods that offers researchers a chance to gain direct experience of social practices in action. It does seem curious, therefore, that the bulk of social and cultural geography in the 1990s was more concerned with how people see and speak about their world than with what they do in it. This is an overgeneralization, of course, but Nigel Thrift (1996) does have a point when he claims that 'academic accounts have not only downgraded the importance of practical activity by trying to represent it as representa-tions ... but may also have understated its power' (p. 33). Such accounts may suffer, too, from putting too much distance between academic reflection, on the one hand, and active engagement with the subjects whose world they are commenting on, on the other.

What approaches are available to us to meet some of these concerns? In his various accounts of the shift towards 'non-representational styles' of working, Thrift (1996, 2000a) identifies many possibilities. Few of us will be well read enough to get through them all! So I shall take just two sets of ideas as a starting point, and I shall raise a few methodological pointers contained in both.

### Pragmatism revisited

The first is a cluster of approaches that I have argued before and that – I continue to believe – hold considerable potential for the practice of qualitative research (Smith 1984). These approaches are rooted in philosophical pragmatism. This set of ideas was developed in North America in the early twentieth century but has undergone something of a renaissance in recent years. Whereas many of the thinkers usually termed 'neo-pragmatists' (Quine, Goodman, Putnam) embrace the linguistic turn from which we now wish to escape, key originators of the idea – especially William James and John Dewey, but also George Herbert Mead – were far more interested in experience, in practice, in what is lived and felt rather than in that which is most clearly conceived. Rorty, the neo-pragmatist most in sympathy with these earlier thinkers, explains: 'Pragmatists insist on nonocular, non-representational ways of describing sensory perception, thought and language, because they would like to break down the distinction between knowing things and using them' (1999: 50). And to quote myself on this topic:

'knowledge is an instrument for action rather than contemplation, and theories are always provisional, never ultimately correct' (Smith 1984: 361).

Pragmatism, then, is a method of knowing through practice. It is about a precarious world, with an uncertain future, at least part of which remains to be made. Being=doing, ideas=action, knowledge is always provisional and the future will always surprise. This is a set of ideas that casts the problem of knowing as a problem of living. It is steeped in the importance of the world as a doing, in which all the old dualisms (structure–agency, reality–appearance, and so on) are replaced by what Rorty calls the 'distinction' (but what I always think of as the point of 'articulation') between pasts and futures. And, crucially, futures are continuously in the making through practices. Pragmatism is therefore both a philosophy and a style of working that turns attention squarely towards the actions and non-actions, the beings and doings, that move the world from the past to the future; it turns attention to the 'now' and to what it is becoming. And it recognizes that this becoming is achieved (not necessarily deliberately) as readily by the small works of 'ordinary' people as by dramatic interventions enacted by states and institutions. Indeed, as I have argued in relation to a variety of local performances, it is here – in the routine goings-on of everyday life as much as through elections or explosive social movements – that the political decisions that drive the world forward are made (Smith 1999).

*The psychic turn*

Pragmatism is one key set of ideas that takes qualitative research beyond the world of representation, into the messy complexity of practice. The other big nudge in this direction comes from psychoanalytical theory. This is something of a minefield to the uninitiated, since it is a hugely controversial area mapped out in complex ways by a difficult and extensive literature. But I am going to be bold, and say that this set of ideas is valuable for practice-oriented qualitative research for at least two key reasons.

First psychoanalysis has a central concern with human subjectivity as it shapes and is shaped by the interface between individuals and the world. Its main project, for Pile (1996: 8–9) is 'to account for the personal meanings that people produce for themselves as they struggle to cope with, and make sense of, the painful realities of everyday life'. Crucially, however, in considering how individuals think and act, a psychoanalytic perspective draws attention to the world of the unconscious. While social science has been reasonably successful in accommodating the subconscious, the psychoanalytic unconscious is quite a different thing: 'It does not, for example, mean not conscious as in yet-to-be-known or taken-for-granted ... [rather] the unconscious is another scene, a parallel process ... it uses its own language, signs and symbols, makes its own connections' (Pile 1996: 76). This unconscious is dynamic and active; it influences thoughts, feelings and actions in the course of daily life. Yet, it is not accessible to subjects, let alone to those studying them.

Now pragmatism emphasizes the partial and provisional character of knowledge, and embraces the fact that all actions have unexpected as well as anticipated outcome. What a psychoanalytic perspective appears to add to this is an element of absolute unknowability – an acknowledgement of the extent to which the human unconscious contributes to an inherently unstable, uncertain and fractured process of knowing in a world whose spaces and places are 'contradictory, multiple, and always becoming' (Rose 2000: 654). What this suggests is that, for the psychoanalyst – for anyone – charting this becoming, it is always necessary to be alert to the presence and the influence of (unconscious) psychic influences as well as (potentially knowable) social and cultural practices.

A second element of psychoanalytic thinking that seems important here is elaborated by Sibley (1995), who – drawing on what the psychoanalysts call 'object relations theory' – talks about the process by which infants' engagement with objects beyond themselves sets up a sense of identity and difference: through emerging boundaries the 'good' and the 'bad' are differentiated. The process of becoming a 'self' therefore goes hand in hand with a setting-up of boundaries during which, at times, as Vogler (2000: 25) puts it, 'the external world [of objects] is taken over by the internal world [of the self] and we respond to others emotionally as if they were disliked parts of our own selves'. This object-relations approach suggests that stereotyping others is part and parcel of defining the self. Practices of exclusion and definitions of self are inextricably bound together. The boundary markings that Sibley refers to – ideas about race categories, gender differences and so on – are therefore shot through with the whole gamut of fears and anxieties that link social difference (differences between selves and others) to distinctions between clean and dirty, ordered and disordered, desire and disgust, and so on. Boundary building is, therefore, about emotional as well as material practices; geographies of inclusion and exclusion are about psychic as well as socio-economic and cultural power relations.

*Putting pragmatism and psychoanalysis into performance*
The work of the pragmatists and the psychoanalysts is immediately helpful on two counts. First, both approaches, and especially that of the pragmatists, stress the importance of engaging as closely as possible with practice. We can never know with certainty, but we *can* engage with people in action. It is interesting, then, that, in stressing how important it is for researchers 'to come much closer to the people whose problems provide the primary justification for the existence of the subject', Sibley (1995: 185) turns to the pragmatist George Herbert Mead for inspiration on the links between the psychic and the bodily.

Second, both styles of working, and especially the psychoanalytic, help us to understand our own reluctance to enter these close engagements. Psychoanalysis helps us appreciate where our own fears, inhibitions and insecurities come from when we are faced with a choice of whether to reflect on texts (which is comparatively easy) or to engage with the action (which is altogether muddier, messier and more uncomfortable). On this second point, Hester Parr's work is instructive. In her sympathetic covert ethnography of the bodily practices of people with mental-health problems (Parr 1998) she notes how, after dancing at a disco with a man from a drop-in centre, 'I left to go to the toilets, and to scrub my hands furiously'. Despite her best intentions, the incident highlighted for her 'the possibility of an invasion of my body, by dirt, fleas or infection'. The solution she chose was not, however, to retreat into the sanitized environment of a more controlled interview setting, but to learn to live with – to work through, to confront, to challenge herself with – the discomfort, unease and threat to self that bodily encounters with (some) subjects might entail.

If we take seriously the emphasis on practice that comes from the pragmatic, the psychoanalytic, and indeed the wider range of what Thrift (1996) calls non-representational styles of working, it requires a rather different cut across the qualitative methods that seem most familiar to us. In-depth interviews, for example, become valuable not for the text into which they are transcribed, but for the conversation as it takes place. So we need to recognize, with Laurier (1999), that talk is different from text; that a story constructed through conversation is different from a story in a book; that what we can know when people talk to one another disappears when that talk is transcribed. Bakhtin (1986) recognized this in distinguishing 'sentences' from 'utterances': the former can be repeated, reproduced and analysed with reference to logic and grammar; the utterance is a practice, knowable only as it is said in one context or another. Picking up on this theme, Laurier (1999) draws inspiration from the ethnomethodologist Harvey Sacks, who studied the 'to and fro' of speech – who regarded speech as a practice that demands attention not just to what is said, but also to how it is told. This in turn is in line with the thinking of neo-pragmatists Richard Rorty and Richard Bernstein, who reflect on the importance of dialogue as a way of challenging the old with the new, the known with the unknown. And this is important for us because, as Laurier argues, there are geographies in this talk: places are made and unmade by the practice of conversing.

But talking is just one aspect of these geographies of the now, and methodologically there may be a need to let go – just a little – of the interview-based approaches that are so popular. I think Edward Said (1991) was moving in this direction when he wrote his book 'musical elaborations', and argued that musicians – musical beings and doings – play a part in how societies tick over or roll on. There is a shift here in his thinking from a preoccupation with representations towards an engagement with 'presentations' – 'with showings and manifestations that characterise social knowledge in use' (Thrift 1996: x). A fascinating example of this shift to presentation is offered by Stacey Holman Jones (1998), in her study of 'women's music' at 'The Club'. Abandoning the usual conventions of ethnographic writing, Holman Jones presents her work as a series of ongoing exchanges – with herself, with the musicians, with her supervisors and mentors – all interspersed with personal musings: 'Turning again, I hear whispers of guitar and piano. I feel the heat of applause and candlelight. These are the things I want to know, the things I can't see or write. I want to feel the artistry, the reverence, the fire of this place' (1998:

153). And so to the presentations, to the events, to the dialogical performances – the acts and the writing of them – that bring the known and the unknowable together in a manuscript that is itself a 'presence' *en route* to a becoming.

Holman-Jones's work gives some clues about the 'how to' of all this, into what a preoccupation with the now means for qualitative research, into how we might engage with practice, with the knowing that consists of doing. Indeed, she adopts the metaphor, and method, that currently seems most widely in vogue – that of performance. It is perhaps not surprising that (in my opinion) recent ethnomusicology provides some of the best examples of this approach. See, for example, the collection edited by Barz and Cooley (1997) for some new ideas on fieldwork; or read Small (1998) on the practice of musicking. There is also a large literature on dance that has become influential in this respect (Thrift 2000a). Working with performance offers one possibility for grasping the present in its own right, for valuing and working with everyday practical activities as they occur, as they make what has been into what is to come. Performance is the articulation of the past with the future, or, as Thrift (2000b: 577) puts it, 'the art of producing the now'.

A performance can be any event, action, item or behaviour. The point is that we are interested in it not (only) for what it represents, but for what it is – a recognition that there is much that can only be known through the doing. Thrift (hopefully tongue in cheek) calls this a 'third kind' of knowledge: 'it is not about knowing what or knowing how, but about knowledge-in-practice' (1996: 17). It is as much about what is done as about what is said; it is a way of recognizing that ways of being include the unknowable and the unsayable.

Performance is most often thought of in relation to geographies of identity, and this largely reflects the key influence of Judith Butler's work (1990, 1993) on feminist thinking within the discipline. Her core point is that identities (and she talks particularly about gender and sexuality) are a doing: they are made and remade through variously regulated processes of repetition. This is often contrasted with foundationalist ideas of identity as inherent, fixed, stable and so on. But this is not what is new about Butler's work: the idea that identity is repeatedly negotiable is as old as social science itself. What is exciting here is the idea of identity as something that can only be accessed as – that only exists in – performance, precisely because it is an expression, an enactment, of what is not 'knowable' (in other ways). As Nelson (1999: 348) argues: 'How individual and collective subjects ... do identity is an inherently unstable and partial process ... it is never transparent because it is always inflected by the unconscious, by repressed desire and difference.' So performance is concerned not with the what but the how; it 'is a living demonstration of skills we have but cannot ever articulate fully in the linguistic domain' (Thrift 1996: 34). Yet these ways of being still include the actions and non-actions, knowings and unknowings, that move the world along.

The only way to get to grips with all this is to engage with it. Thrift (2000c) quips that what is required is not participant observation in the traditional sense, but observant participation. What I think he means by this is to come close to subjects, to be with subjects in a way that enables us to recognize how the various 'skills' that keep the world moving are acquired and implemented. This might be by *being* a performer (which everyone is, in some sense); alternatively, clues might be gleaned from what those already skilled in performance – artists – do. This seems to be what lies at the core of Kotarba's 'synthetic performance ethnography' (1998), in which the researcher is both producer and analyst of particular performance events. And when Thrift (2000a) discusses dance, rather than doing it (necessarily), he seems to be arguing that appreciating (watching, hearing, sensing, experiencing) the performative acts that dancers engage in can alert us to the performative acts that constitute many other realms: the life of cities, the work of institutions and so on. Crucially, though, whether by doing it or being part of a performance event (metaphorically, in the 'audience'), working with performance is working with a conception of knowing and being that goes beyond the visual, the textual and the linguistic.

Indeed what has drawn me to performance has been its relevance to conceptualizing, creating and occupying not just 'the usual' political and economic spaces, but also emotional spaces – feelings and sensings that are both knowable and unknowable, but whose significance in shaping the world and its future is definitely underplayed. Here there is a clear steer from psychoanalysis in which 'human beings are seen as being linked to each other by a profound emotional intersubjectivity or emotional communication in which they continually give and receive emotional messages, although this is usually unconscious and not recognised by those involved' (Vogler 2000: 23). So (while I think Vogler overstates the extent to which emotional relations take place in the unconscious), I have written about musical performance as a way of claiming emotional space: 'working in a way that words cannot grasp, embodied music [i.e. music in performance] creates an emotional space powerful enough to displace (replace) the speech act as a means of communication – as a means, for example, of experiencing and negotiating place and identity' (Smith 2000: 632).

Performance, then, is about the way identities and spaces are made and remade – about how they are enacted at the moment – through conscious and unconscious, psychic and material processes. It is also about the way things change, or do not change (the latter being as active a process as the former). It is about the prospect for the new, precisely because there is no inevitability (even if there is a tendency) for repetition to equate with reproduction. Performance is improvised rather than (pre)scripted. Therefore, the world has an openness as well as an unknowability. Things could have been different; and they might be again. The world is a doing and a making that may be regulated by the past, that may be constrained in its present, but that remains open to the powers of imagination and invention – charged as they are with emotion as well as intention, with subjectivity instead of certainty, with pleasure, pain, power and politics.

## Conclusion

I have used this chapter first to suggest that choosing qualitative methods is a political rather than a philosophical issue, and to argue that this political dimension remains under-explored and under-exploited. This partly reflects a just concern with the ethical dilemmas of qualitative research, but it also expresses an ongoing and frustrating reluctance on the part of researchers to grasp the nettle, in a world where no act (including not acting) is value free.

Next, I noted a tendency for researchers today to harness particular qualitative methods – more precisely, particular ways of using certain methods and the results they yield – to particular ways of conceiving of the world. Here I distinguished between a concern with texts and representations, on the one hand, and an interest in beings and doings – with practices – on the other. I do not regard these traditions as mutually exclusive approaches to knowledge. Indeed, Nash (forthcoming) warns against the temptation to treat them in this way. The world *is* like a book, but by now the pages are well thumbed. So, moving beyond representations and interpretations – reconsidering the way the world's a stage – is probably the most intriguing turn in recent qualitative research.

There are 'non-representational' strains of working in many schools of thought. However, I have suggested a combination of pragmatism and psychoanalysis as an appealing starting point for studying partially knowable worlds that are actively in the making through a combination of material and psychic, socio-economic and emotional practices. I conclude with the thought that an appropriate conceptual and methodological tool for this kind of qualitative research can be shaped from the practice of performance.

## Key references

Bailey, C., White, C. and Pain, R. 1999: Evaluating qualitative research: dealing with the tension between 'science' and 'creativity'. *Area* 31(2), 169–78.

Baxter, J. and Eyles, J. 1997: Evaluating qualitative research in social geography: establishing 'rigour' in interview analysis. *Transactions of the Institute of British Geographers* NS 22(4), 505–25.

Laurier, E. 1999: Geographies of talk: 'Max left a message for you'. *Area* 31, 36–45.

Parr, H. 1998: Mental health, the body and ethnography. *Area* 30, 28–37.

Smith, S. J. 2000: Performing the (sound)world. *Environment and Planning D: Society and Space* 18(5), 615–38.

Thrift, N. 2000a: Afterwords. *Environment and Planning D: Society and Space* 18, 213–56.

## References

Back, L. and Solomos, J. 1993: Doing research, writing politics: the dilemmas of political intervention in research on racism. *Economy and Society* 22, 178–99.

Bailey, C., White, C. and Pain, R. 1999: Evaluating qualitative research: dealing with the tension between 'science' and 'creativity'. *Area* 31(2), 169–78.

Bakhtin, M. 1986: *Speech genres and other late essays*, trans. V. W. Mcgee, eds. C. Emerson and M. Holquits. Austin, Tex.: University of Texas Press.

Barthes, R. 1977: The death of the author. In Barthes, R. (ed.), *Image, music, text*. London: Fontana, 142–8.

Barz, G. F. and Cooley, T. J. (eds.) 1997: *Shadows in the field: new perspectives on fieldwork in ethnomusicology*. New York and Oxford: Oxford University Press.

Baxter, J. and Eyles, J. 1997: Evaluating qualitative research in social geography: establishing 'rigour' in interview analysis. *Transactions of the Institute of British Geographers* NS 22(4), 505–25.

Bowes, A. 1996: Evaluating and empowering research strategy: reflections on action-research with South Asian women. *Sociological Research Online* 1(1) <http://www.socresonline.org.uk/socresonline/1/1/1.html>.

Burgess, J., Limb, M. and Harrison, C. M. 1988: Exploring environmental values through the medium of small groups: 1. Theory and practice. *Environment and Planning A* 20, 309–26.

Butler, J. 1990: *Gender trouble: feminism and the subversion of identity*. New York: Routledge.

—— 1993: *Bodies that matter: on the discursive limits of 'sex'*. New York: Routledge.

Cresswell, J. W. 1998: *Qualitative inquiry and research design. Choosing among five traditions*. London: Sage.

Duncan, J. 1990: *The city as text: the politics of landscape interpretation in the Kandyan kingdom*. Cambridge: Cambridge University Press.

Dyck, I. 1999: Using qualitative methods in medical geography: deconstructive moments in a subdiscipline? *Professional Geographer* 51(2), 243–53.

Holman-Jones, S. 1998: Kaleidoscope notes: writing women's music and organizational culture. *Qualitative Inquiry* 4, 148–77.

Howes, D. 1991: Sense and non-sense in contemporary ethno/graphic practice and theory. *Culture* 11, 65–76.

Johnson, A. 1996: 'It's good to talk': the focus group and the sociological imagination. *Sociological Review* 44, 517–38.

Katz, C. 1992: All the world is staged: intellectuals and the project of ethnography. *Environment and Planning D: Society and Space* 10, 495–510.

Keith, M. 1992: Angry writing: (re)presenting the unethical work of the ethnographer. *Environment and Planning D: Society and Space* 10, 551–68.

Kotarba, J. A. 1998: Black men, black voices: the role of the producer in synthetic performance ethnography. *Qualitative Inquiry* 4, 389–404.

Laurier, E. 1999: Geographies of talk: 'Max left a message for you'. *Area* 31, 36–45.

Morgan, D. 1996: Focus groups. *Annual Review of Sociology* 22, 129–52.

Nash, C. (forthcoming) Performativity in practice: some recent work in cultural geography. *Progress in Human Geography*.

Nelson, L. 1999: Bodies (and spaces) do matter: the limits of performativity. *Gender, Place and Culture* 6, 331–53.

Offe, C. 1984: *Contradictions of the welfare state*. London: Hutchinson.

Parr, H. 1998: Mental health, the body and ethnography. *Area* 30(1), 28–37.

Pile, S. 1996: *The body and the city: psychoanalysis, space and subjectivity*. London: Routledge.

Ricoeur, P. 1971: The model of the text: meaningful action considered as text. *Social Research* 38, 529–62.

Roberts, H., Smith, S. J. and Bryce, C. 1995: *Children at risk?* Buckingham: Open University Press.

—— —— and Lloyd, M. 1993: Prevention is better . . . *Sociology of Health and Illness* 15, 447–63.

Rorty, R. 1999: *Philosophy and social hope*. London: Penguin Books.

Rose, G. 2000: Psychoanalytic theory, geography and. In Johnston, R., Gregory, D., Pratt, G. and Watts, M. (eds.), *Dictionary of human geography*. Oxford: Blackwell, 653–5.

Said, E. 1991: *Musical elaborations*. London: Chatto & Windus.

Sibley, D. 1995: *Geographies of exclusion*. London: Routledge.

Silverman, D. 2000: *Doing qualitative research: a practical handbook*. London: Sage.

Small, C. 1998: *Musicking: the meanings of performance and listening*. Hanover: University Press of New England.

Smith, S. J. 1981: Humanistic method in contemporary social geography. *Area* 13, 193–8.

—— 1984: Practising humanistic geography. *Annals of the Association of American Geographers* 74, 353–74.

—— 1999: The cultural politics of difference. In Massey, D., Allen, J. and Sarre, P. (eds.), *Human geography today*. Cambridge: Polity, 129–50.

—— 2000: Performing the (sound)world. *Environment and Planning D: Society and Space* 18(5), 615–38.

Thrift, N. 1996: *Spatial formations*. London: Sage.

—— 2000a: Afterwords. *Environment and Planning D: Society and Space* 18, 213–56.

—— 2000b: Performance. In Johnston, R., Gregory, D., Pratt, G. and Watts, M. (eds.), *Dictionary of human geography*. Oxford: Blackwell, 577.

—— 2000c: Non-representational theory. In Johnston, R., Gregory, D., Pratt, G. and Watts, M. (eds.), *Dictionary of human geography*. Oxford: Blackwell, 556.

Vogler, C. 2000: Social identity and emotion: the meeting of psychoanalysis and sociology. *The Sociological Review* 48, 19–42.

# 3

# At the drawing board: developing a research design

## Gill Valentine

## Introduction

The term 'qualitative methods' covers a confusing array of alternative research techniques, including those such as interviewing, participant observation and focus groups, which are outlined in the chapters of this book, and other less common methods such as diaries and auto-photography (see e.g. Aitken and Wingate 1993). This chapter reflects on how to design a research project and choose the most appropriate method or combination of methods to use.

A research design is a result of a series of decisions we make that emerge from our knowledge of the academic literature, the research questions we want to ask, our conceptual framework and our knowledge of the advantages and disadvantages of different techniques. In this chapter I therefore frame the discussion of research design in terms of four key decision-making processes: first about what research questions to ask, second about the most appropriate method(s) to employ, third about the practicalities of doing fieldwork and fourth in relation to ethical considerations. Although these decision-making points are presented here in a linear way, it is important to remember that in practice these issues are more complexly bound up with each other and that because of this these decisions are often made simultaneously or in conjunction with each other. I conclude the chapter by outlining an example of how a research design was developed for one of the projects on which I have worked.

## Decision 1: what are your research questions?

Before you jump into the deep end by starting to think about which methods to use, it is important first to have identified what your research questions are and therefore what is the substantive focus of your project, because the nature of the questions that you want to ask will shape the most appropriate ways of investigating them. Gatrell (1997) makes a number of useful suggestions about how to choose and define a topic.

Research topics and the specific questions they generate can often begin with your own observations. You may have been particularly interested in something you heard in a lecture or read in a journal or a newspaper; you may have noticed that your everyday personal experiences do not fit in with academic theories or research findings that you have read; you may be questioning why particular theories or ideas apply to certain groups of people and not others; why particular groups of people use certain spaces, or how national policies or issues on the public agenda are being played out in a specific place. For example, when I was an undergraduate I took a course on environmental perception. While some of the lectures and articles on the reading list focused on people's attachment to place, some of the other women in the class and I began to talk about how our own feelings about, and use of, particular spaces within the city were shaped by our fear of violence. As a result I tried to read up on this but found that at the time, although there was some feminist work on women and the built environment, there was no geographical literature on this issue. The work I found on women's fear of crime in other disciplines such as sociology and criminology paid little attention to the spatial dimensions and implications of violence. This led me to ask a series of questions about how women use and experience public space (see Valentine 1989, 1992).

Once you have reflected on what questions are pressing to you or bother or intrigue you, it is important to carry out bibliographic searches to find how other researchers have dealt with your topic. To get started use your reading lists, look for reviews of your chosen field in the journal *Progress in Human Geography* and talk to a tutor. Then search more methodically by using *GeoAbstracts*, which provides an index of the abstracts that are usually printed at the start of journal articles summarizing their contents, and citation indexes, which are available online. These list all the other articles that have cited a reference and so are particularly helpful for enabling you to trace how work has developed on your topic area since a key paper was written. Other computerized databases can also enable you to search for articles using key words, and do not forget to search the Internet.

A literature review serves several different roles. It embeds the research questions in broader empirical traditions and in doing so provides evidence of the significance of your project and its contribution to knowledge. In other words, it validates what you want to do. It also enables you to define the intellectual traditions that have been drawn on in the study of your topic, so helping you to define key terms of reference and to develop a conceptual framework that can guide your research. It is particularly important to be aware of the philosophical approach (for example, positivism, humanism, Marxism, postmodernism, and so on) adopted by different researchers to your project, because the approach that is taken shapes what research questions are understood as valid, what methods are regarded as appropriate and what sort of knowledge is produced (Graham 1997). The main philosophies underpinning human geography are outlined in Cloke *et al.* (1991).

A good literature review is one that not only summarizes all the major studies in your chosen field of research, but that is also critical of the way that the topic has been addressed in the past, identifying both the strengths and weaknesses of previous work and the gaps within it. On the basis of this, rather than merely replicating previous studies, your project should aim to refine or build on the literature. This might be by asking different questions, employing different methods, approaching the topic from a different philosophical perspective, or exploring the issues with a different group of people or in a different geographical context.

On the basis of your own thinking about the topic, having grounded it in the relevant theoretical and empirical literatures (and if possible having discussed it with other students and your tutor), you need to move towards framing your specific research questions. These might include questions about what discourses (or sets of beliefs) you can identify, what patterns of behaviour/activity you can determine, what events, beliefs and attitudes are shaping people's actions, who is affected by the issue under consideration and in what ways, and how social relations are played out, for example. It is important to have a strong focus to your research questions, rather than adopting a scatter-gun approach asking a diverse range of unconnected questions. You must also bear in mind the time and resource constraints on your research, which ultimately define the range and scale of what you can achieve (see Decision 3 below).

Although you should begin your research with a clear set of research questions, this is not to suggest that they cannot or should not change as the study develops. The way research is written up in academic journals often represents it as a linear, pristine, ordered process. Yet, in practice, most projects are actually more messy, frustrating and complex. For example, unanticipated themes can emerge during the course of fieldwork that redefine the relevance of different research questions. Likewise, access or other practical problems can prevent some research aims being fulfilled and lead to a shift in the focus of the work. Therefore, you should be aware that your research questions may evolve during the course of your project.

## Decision 2: what research method(s) should you adopt to address these questions?

The choice of research methods usually flows conceptually and logically from the research questions. In other words, you should have a rationale for your choice of methods, rather than just thinking: *everyone uses interviews so that is what I should do*, or *participant observation looks easy so that is my method*. While qualitative techniques emphasize quality, depth, richness and understanding, instead of the statistical representativeness and scientific rigour that are associated with quantitative techniques, this does not mean that they can be used without any thought. Rather, they should be approached in as rigorous a way as quantitative techniques.

There is no set recipe for research design; different methods have particular strengths and collect different forms of empirical material. The most appropriate method(s) for your research will therefore depend on the questions you want to ask and the sort of information you want to generate. A sound research design needs a logical choice of methods that meets the aims set and generates data in a way that the researcher can handle and interpret. In order to appreciate this you need to have an understanding of a range of different qualitative methods and the sorts of materials they produce.

In-depth interviews are used to get participants to provide an account of their experiences, of how they view their own world and the meanings they ascribe to it (see Part 2). They vary in style from very conversational to more formal. The advantage of interviewing is that it can generate a lot of information very quickly; it enables the researcher to cover a wide variety of topics, to clarify issues raised by the participant and to follow up unanticipated themes that arise. The disadvantage of this method is that it largely depends on the interpersonal and listening skills of the interviewer. Interviewers may not ask the right or appropriate questions or they may not be understood by an interviewee; interviewees may not be willing to share their experiences, particularly about sensitive or personal topics. Ultimately, interviews only generate information about what the participants say they do rather than their actual practices.

Focus groups or in-depth groups (see Part 3) share similar advantages to interviewing but have the added benefit of enabling the researcher to explore how meanings and experiences are negotiated and contested between participants. This, however, also brings added problems, in that the group facilitator needs to be able to manage the relationships between group members (see Burgess *et al.* 1988a, b; Holbrook and Jackson 1996; Longhurst 1996).

Participant observation (see Part 4) is a technique that involves living, working or spending periods of time in a particular 'community' in order to understand people's experiences in the context of their everyday lives. As the name implies, the researcher participates in and observes the 'community', making notes about events, activities and the behaviour of, and relationships between, its members (Cook 1997). The nature of the researcher's role can vary in terms of whether the study is overt or covert; and according to the extent to which the researcher moves from the role of observer to participant.

The advantage of participant observation is that it produces rich detail and description (especially non-verbal information) about people in the context of their everyday lives. As such the researcher may observe things that people would not talk about in an interview (for example, because they are taken so much for granted that the respondent does not think they are important or interesting, or because something is sensitive and therefore difficult to discuss). Whereas an interview provides information from only one person's perspective, participant observation enables the researcher to gain a broader perspective or overview of a 'community' and the relationships within it. Like the other methods, this technique also has limitations. It is often difficult to negotiate access to, and it takes time to become properly immersed in, 'communities'. As researchers we also have much less control over what information we collect than in an interview situation, and may feel swamped by all the different themes and perspectives that might emerge. This can make defining the boundaries of a study and its focus difficult.

Finally, other methods such as auto-photography, video or audio diaries, drawing or map making are all techniques that give our respondents the opportunity to present their own accounts of their lives in ways that are not framed by our questions as researchers and that do not rely on their verbal or literacy skills or our ability to develop a rapport with them. The understanding or interpretation of material generated in these ways can also, of course, be explored or validated by employing more traditional methods such as interviewing participants about what they have filmed, recorded or drawn and why. While these sort of techniques do not necessarily draw on people's verbal or written ability, they do require other skills – for example, a basic level of technical expertise – and as such some participants are intimidated by them. Filming is also potentially intrusive of individuals' or households' privacy, and because of the usual need to respect participants' confidentiality it is very hard to use these sorts of visual materials in written work or presentations without compromising their anonymity. Finally, the cost of buying/borrowing equipment, insurance, films, and developing or editing can be prohibitively high.

In the process of research design it is important not to view each of these methods as an either/or choice. Rather, it is possible and often desirable to mix methods. This process of drawing on different sources or perspectives is known as *triangulation*. The term comes from surveying, where it describes using different bearings to give the correct position. In the same way as surveyors, as researchers we can use multiple methods or different sources of information to try and maximize our understanding of a research question. These might be both qualitative and quantitative (see e.g. Sporton 1999).

For example, if you were studying a motorway protest group (see Routledge 1997), you might begin by carrying out participant observation to gain an understanding of the participants and their group dynamics, but then use individual interviews as a way of exploring some individual motivations for participating and some of the tensions between members of this 'community'. Finally, you might want to ask some or all of the group to make a video or take some photographs that capture their feelings about the motorway. When multiple methods are used in this way, the material generated by each technique does not always reinforce or validate that produced by the others, but rather different techniques may throw up apparently very contradictory findings. For example, the participant observation of the motorway group may paint a picture of a very political group, divided into different factions, yet the video diary they produce might present a very coherent, united front, while the interviews might expose the very different motivations of individuals for being there (some of which may have nothing to do with the road but might reflect personal or emotional problems at home) and reveal that 'behind-the-scenes' the relationships between individuals might not accord with what you have observed.

Such findings are not a problem but rather show how successful you have been at capturing the complexities, contradictions, ambiguities and messiness of human behaviour and everyday life. The secret of a good research design is not, therefore, to produce a nice linear, consistent story, but rather to be able to recognize why the accounts generated by particular methods can be so different and to be able to use these differences to produce a fuller interpretation and understanding of your research questions.

If you plan to mix your methods, it is important to think about how the different techniques you plan to use will be integrated both in the course of the research and in terms of the way you use the data and write up the findings. Different techniques should each contribute something unique to the project (perhaps addressing a different research question or collecting a new type of data) rather than merely being repetitive of each other. I illustrate how a combination of methods were used to answer different research questions in my final example.

## Decision 3: what are the practicalities of your fieldwork?

It is often relatively easy to come up with a range of research questions, but the real trick is defining those that are actually doable within the time, geographical and financial constraints of a student research project. The nitty-gritty practicalities of who, what, when, where, for how long inevitably shape the choices we can make about our aims and methods.

It is important to bear in mind that the research that is written up by academics in journals and books is often conducted over several years and is commonly funded by substantial grants. As such, the scale of this sort of research is very different from the scale at which student research projects must be pitched. It is not possible to replicate or fully develop all the objectives of a two-year piece of academic research in a three-month student dissertation or project. Rather, it is often best to begin by identifying the limitations of your proposed study and recognizing what you will and will not be able to say at the end of it.

Any choice of methods leads to the question of how to gain access to potential participants or 'communities' and decisions about the number of people or organizations that need to be recruited. Again, this often involves juggling academic criteria and ideals against the practicalities of what it is possible to achieve in a limited time span. While it is important to gain a range of views in order to achieve as full an understanding as possible of your research questions, qualitative research does not aim to be statistically representative, and so, unlike in quantitative studies, it will be the depth and richness of your encounters rather than the number of people who participate in the study that matter. To this end, qualitative researchers usually employ what is known as illustrative sampling, rather than the random sampling techniques used to select a representative sample of a population adopted by those using questionnaires. In this way, choosing who to recruit for interviews, focus-group discussions or audio-visual work is usually a theoretically motivated decision in which researchers draw on their understandings of the issues to decide which angle or perspective they need to explore (for example, whether to recruit men and women, or people of different ages, and whether their place of residence or work matters and so on). As part of this it is also important to reflect on your own identity and how this will influence or shape the interactions you might have with those with whom you want to work. This self-critical introspection is what academics mean when they talk about identifying your *positionality* or recognizing your *reflexivity* (see e.g. England 1994; WGSG 1997; see further Kneale, this volume).

Recruiting participants to take part in research projects often takes longer than you anticipate. Some groups are particularly difficult to gain access to. For example, because the elderly are a group who feel vulnerable about their personal safety, they are understandably often reluctant to allow strangers (that is, researchers) into their homes. Similar popular anxieties about children's safety and the demands of school timetables and out-of-school activities (as well as legal considerations) can mean that it is difficult to negotiate access to work with young people. Marginal groups such as protesters or political activists, lesbian, gay or bisexual organizations or disabled support groups may be wary of 'outsiders' asking sensitive questions and of the purpose of the research, and may challenge your motivation and ethics. The main obstacle to recruiting participants from commercial organizations is often the gatekeeper, usually a secretary, who controls access to senior employees. Business people are commonly accountable for their time and so are often reluctant to respond to student queries or will consent to only the briefest of interviews (for discussions of studying elites, see Schoenberger 1991; McDowell 1992; or Herz and Imber 1995).

Gaining access to a research community or recruiting individuals is crucial to the viability of any study. It is important, therefore, when designing a project to be realistic about what it will be possible to achieve, to plan carefully how you will introduce yourself to those concerned, and to allow plenty of time to do so. It is also worth anticipating what you will do if your recruitment strategies fail and having a contingency plan in place. For example, if you are studying place promotion in advertising, your research questions may focus on how several chosen advertisements are produced by organizations and advertisers and therefore you may plan to interview art directors at key advertising agencies. This is a risky research design, however, because advertising executives are notoriously difficult to contact. A fallback position might be to allow yourself two weeks to recruit these professionals; if you are unsuccessful you might shift the focus of your research questions from how these advertisements are produced towards how these images are consumed by particular social groups. This might in turn entail a shift in your research methodology away from interviews with elite individuals towards focus groups with members of particular 'communities of practice' (for example, students or families).

When you are estimating the number of interviews or focus groups you need to conduct, be realistic about what it is possible to achieve within your time frame. Most people have busy lives and so it can be difficult for them to make time to talk to you, and they will not necessarily be available when you want to speak to them. At the same time, do not forget when you are planning your project that you will also have other demands on your time, and to include your holidays, other work and leisure commitments and travel time (and costs) into your research schedule. Remember that doing qualitative work requires a lot of concentration and mental energy and as such is both stressful and tiring, so there is a limit to how much you can achieve in the field in any one day. Other practicalities such as the availability of tape recorders, cameras, transcribers or access to transport can also define the parameters of your project. Drawing up a time-management chart or work schedule at the research-design stage can be an effective way of working out how much you can achieve in your study, and later on can also serve as a useful indicator if you are slipping behind.

The time it takes to write up a field diary or transcribe an interview or focus-group discussion (if you are a competent typist it takes about four hours to transcribe one hour of tape, much longer if your typing is slow or the sound quality on your tape is poor) is often a forgotten element of research design. There is no point conducting lots of interviews and focus groups if you do not have the time to transcribe and analyse the material. It is better to carry out less fieldwork in more depth and develop a detailed analysis and genuine understanding of that material than to collect a large body of data that you cannot get to grips with and of which you can produce only a cursory interpretation.

Thinking about the way you intend to analyse the information you collect using qualitative methods should form an integral part of any research design. Part 5 of this book contains several chapters that look at different interpretative strategies. It is worth reading these chapters at the design stage, as this will help you to develop your methodology and ensure that you collect your material in a way that you can analyse (see also Crang 1997). This issue has become even more relevant with the recent emergence of a range of computer software packages for analysing qualitative materials, some of which require the data to be input in different forms (Miles and Huberman 1994). Knowing in advance how you intend to interpret and use your data is particularly important if you want to use unconventional methodologies such as auto-photography or video diaries where there are less clearly established interpretative strategies.

Many of the issues and questions outlined in this section can be resolved by carrying out a pilot study before your actual research begins. By conducting some preliminary interviews or focus groups (perhaps with people you know who can give you feedback on your technique as an interviewer or facilitator) you can get a feel for how much you can cover in one meeting, what sort of questions or issues prompt your participants to talk and which are confusing or need clarification, and what sort of material you might generate in this way. Pilot studies also make you think about and practise how to introduce both yourself and what you are doing and enable you to become familiar with using a tape recorder or video camera.

While planning ahead (and in doing so drawing on the experience of your tutor and other researchers) is crucial to developing an effective research design, it is also important to remember that you should always remain flexible. Unexpected problems (especially in relation to access) often crop up; likewise one of the strengths of qualitative research is that unanticipated issues can emerge or become significant during the course of a project.

# Decision 4: what ethical issues do you need to consider?

An awareness of the ethical issues that are embedded in your proposed research questions and possible methodologies must underpin your final decisions about the research design. The most common ethical dilemmas focus around: participation, consent, confidentiality/safeguarding personal information and giving something back to those who have participated in the research (Alderson 1995; Valentine 1999a).

One problem with approaching potential research participants (either directly or through a third party) is that people (particularly children or the elderly) often find it difficult to say 'no' (especially if, in the case of children, you have an assumed authority over them as an adult). It is important, therefore, when you approach informants that you offer them the option of opting into your research rather than phrasing your request in a way that implies or assumes that you expect a 'yes' answer and puts them into the uncomfortable position of having to refuse you. When working with children or 'vulnerable groups', the sociologist Priscilla Alderson (1995) suggests that it is a good idea to encourage them to practise saying 'no' before they are asked to take part in the study. If your research involves more than one stage, the participants should always be asked to opt into each stage of the research separately rather than be put into the more difficult position of telling you that they want to opt out in the middle of the project.

If you are using qualitative methods such as interviewing or focus groups (though obviously not covert participant observation), you should always obtain the informed consent of those to whom you talk. This means providing them with written details about the purpose of your study and an agreement for them to sign that states that they are willing to participate in the project and giving their consent for you to use what they say in an anonymous form when writing up your findings. This ensures that the respondents know what they are committing themselves to and that you have the necessary permission to use the data generated by the study. Under UK law (although this may not be true elsewhere) the copyright in relation to quotations in an interview transcript belongs not to the researcher who recorded and typed up the material but rather to the person who spoke the words. As such, if researchers use quotations from transcripts without this form of consent, they are technically breaching copyright law.

Confidentiality is a particularly thorny issue. In many research projects you will be asking informants to talk to you either about personal experiences or opinions – often things that are usually considered 'private' issues and are only ever discussed with family and friends, if at all – or about professional matters that might include commercial, political or potentially sensitive information. If particular excerpts from these conversations are made known to other household members, colleagues or the wider 'community' they may have serious personal or professional consequences for those concerned or at least cause them some embarrassment or discomfort. It is important, therefore, that you maintain the confidence of those to whom you talk. This can affect the organization and timing of your empirical work. For example, it might mean that you need to arrange to interview different household members separately rather than together, which will greatly increase the amount of fieldwork you need to do (Valentine 1999b), or that you need to interview particular employees in a certain order, or without the knowledge of their colleagues, which might shape the timing and location of your research.

In order to safeguard personal information, participants should also be allocated pseudonyms. These should be applied both to your interview transcripts and when writing up your findings, so that the anonymity of your respondents is protected on both a day-to-day basis and in the process of presenting your results.

Finally, as researchers we usually ask our respondents to give us something (their time, experiences, thoughts and even emotions) for nothing, and even sometimes at financial or personal cost to themselves. The least we can do in return therefore is to feed our findings back to them. This might be in the form of a brief outline report summarizing the key results. Often, because of the sort of issues about confidentiality and the sensitive nature of some material outlined, it might be necessary to write targeted reports for different audiences. These commitments, therefore, also need to be built into your research design and the time frame of your study.

Thus, while ethical issues may seem routine or moral questions rather than anything that is intrinsic to the design of a research project, in practice they actually underpin what we do. They can shape how we approach gatekeepers or go about seeking access to respondents, what questions we can ask, who we talk to, and where, when and in what order. These choices in turn may have consequences for what sort of material we collect, how it can be analysed and used, and what we do with it when the project is at an end. As such, ethics are not a politically correct add-on but should always be at the heart of any research design.

## Putting it all together: an example of a research design

The issues identified in the previous sections are perhaps best illustrated through the example of a research project I carried out with Sarah Holloway about children's use of the Internet (Holloway et al. 1999; Holloway and Valentine 2000). We became interested in this topic as a result of reading the British media coverage of children and information communication technologies (ICT). These appeared to us to contain paradoxical representations of childhood. On the one hand, unlike most other understandings of child–adult relations, discourses in newspaper articles assumed children to be equally technologically competent as adults, if not more so. On the other hand, children's very competence at using the Internet was also alleged to be putting them at risk of corruption (for example, through their ability to access pornography or other unsuitable materials) or abuse (for example, through contacts made on-line). It was also being blamed for causing children to stay indoors playing computer games rather than engaging in what were represented as more imaginative and social forms of outdoor play. The popular press also suggested that ICT might be undermining children's social skills, friendships and relations between adults and children within the home.

As part of our literature review we brought together three different bodies of work. From the children's geographies literature, we were able to take writings about young people's use of public space and to use these to inform our ideas about children's use of cyberspace. From the sociology of childhood literature, we focused on sociologists' understandings of children's competence and agency; and from the social studies of technology literature, we drew on the theoretical perspective of actor network theory. At the same time, we also recognized that each of these literatures had its own limitations. Geographers have not paid enough attention to children's accounts of their own experiences, sociologists have ignored the importance of the spatiality of children's lives, while the technology literature has predominantly focused on techno-addicts rather than the more mundane everyday ways computers and the Internet are used by the less technologically literate.

Our understandings of both the popular debates about children and the Internet and our reading of these different academic literatures helped us to develop a series of research questions. Specifically, we aimed to collect new information about children's use of ICT and the way it fits into their everyday lives by asking questions about: what information and communication resources are being used by children; the role of ICT in household relationships; the way that ICT might be affecting the constitution of children's friendship networks and their sense of place in the world from a local to a global scale; how ICT is impacting on the structure and organization of children's time and whether these technologies are enabling children to renegotiate the artificial boundaries of adulthood (for example, by passing as someone of a different age when on-line).

In order to address these questions we chose to mix our methods. A large-scale questionnaire survey and time–space diaries distributed through schools were used to obtain factual information about children's access to, and use of, ICT, both at school and home. Participant observation was then carried out in selected technology lessons. This enabled us to see what children were actually doing and to understand the ways that computers were used in their interactions with each other and with their teachers. From these observations we were then able to identify the different friendship groups within the classroom. Focus groups were then carried out with each of these friendship groups. These enabled us to ask questions about some of our observations of the use of ICT at school and within these peer groups. These discussions also enabled us to explore how the meanings of ICT were negotiated and contested between children.

The information we collected from the survey and the focus groups helped us to identify a range of children, including those with different levels of access to technology at home, those with different levels of competence in using a networked PC and those for whom ICT held a range of different meanings. We then asked these children whether they would be willing to be interviewed individually at home and whether they would give us permission to contact their parents to ask them whether they would also participate in the study as part of the household phase of the research. The interviews with children and parents, which were conducted separately (Valentine 1999b), allowed us to collect information about the role of ICT within the household, and its impact on children's use of time and on relationships between adults and children. These interviews also exposed differences between adults' and children's views, tensions within households, some of the ways that adults regulate young people's on-line activities and how these are contested and subverted.

During the course of the work we were surprised that we found it difficult to contact many children who were techno-enthusiasts and who used the Internet extensively. In order to address our research questions about children's on-line activities, we decided to adapt our research design by using e-mail and the Internet to recruit additional children (not from our case-study schools) to be interviewed specifically about their use of cyberspace.

The whole research design was informed by our awareness of the ethical questions concerning children's participation in research projects (Alderson 1995; Valentine 1999a). Specifically, children were asked to opt into the research rather than to opt out, both children and parents were asked for their consent to participate in the study, and each respondent was allocated a pseudonym. When the research was completed, all the participants were sent summaries of the findings. Targeted reports of the results were also sent to relevant government, education and public bodies, and a press release was issued that led to some of the findings being disseminated on television and in the national press.

## Conclusion

In this chapter I have outlined the decision-making processes involved in developing a research design, although I have stressed that this is not a straightforward or linear process. I have argued that during the research-design process you will make important decisions about theoretical and empirical context, methods, practicalities and ethics that largely determine the conduct and outcomes of your research project. Yet I have also emphasized that there needs to be flexibility in the research process, which allows researchers to modify their decisions and so respond to unforeseen events and opportunities. It is important, however, that you keep a record of all the decisions you make about your research as it evolves, and of the rationale behind them, so that the research-design process is made transparent and can be critically evaluated.

## Key references

Crang, M. 1997: Analysing qualitative materials. In Flowerdew, R. and Martin, D. (eds.), *Methods in Human Geography*. London: Longman.

Gatrell, A. C. 1997: Choosing a topic. In Flowerdew, R. and Martin, D. (eds.), *Methods in Human Geography*. London: Longman.

Sporton, D. 1999: Mixing methods of fertility research. *Professional Geographer* 51, 68–76.

Valentine, G. 1999a: Being seen and heard? The ethical complexities of working with children and young people at home and at school. *Ethics, Place and Environment* 2, 141–55.

# References

Aitken, S. C. and Wingate, J. 1993: A preliminary study of the self-directed photography of middle-class, homeless and mobility-impaired children. *Professional Geographer* 16, 553–62.

Alderson, P. 1995: *Listening to children: children, ethics and social research*. Ilford: Barnardos.

Burgess, J., Limb, M. and Harrison, C. M. 1988a: Exploring environmental values through the medium of small groups: 1. Theory and practice. *Environment and Planning A* 20, 309–26.

—— 1988b: Exploring environmental values through the medium of small groups: 2. Illustrations of a group at work. *Environment and Planning A* 20, 457–76.

Cloke, P., Philo, C. and Sadler, D. 1991: *Approaching human geography*. London: Paul Chapman.

Cook, I. 1997: Participant observation. In Flowerdew, R. and Martin, D. (eds.), *Methods in human geography: a guide for students doing a research project*. London: Longman, 127–50.

Crang, M. 1997: Analysing qualitative materials. In Flowerdew, R. and Martin, D. (eds.), *Methods in human geography: a guide for students doing a research project*. London: Longman, 183–96.

England, K. V. L. 1994: Getting personal: reflexivity, positionality and feminist research. *Professional Geographer* 46(1), 80–9.

Gatrell, A. C. 1997: Choosing a topic. In Flowerdew, R. and Martin, D. (eds.), *Methods in human geography: a guide for students doing a research project*. London: Longman, 36–45.

Graham, E. 1997: Philosophies underlying human geography research. In Flowerdew, R. and Martin, D. (eds.), *Methods in human geography: a guide for students doing a research project*. London: Longman, 6–30.

Herz, R. and Imber, J. 1995: *Studying elites using qualitative methods*. London: Sage.

Holbrook, B. and Jackson, P. 1996: Shopping around: focus group research in north London. *Area* 28(2), 136–42.

Holloway, S. L. and Valentine, G. 2000: Corked hats and Coronation Street: British and New Zealand children's imaginative geographies of the other. *Childhood* 7, 335–57.

—— —— and Bingham, N. 1999: Institutionalising technologies: masculinities, femininities and the heterosexual economy of the IT classroom. *Environment and Planning A* 32, 617–33.

Longhurst, R. (1996): Refocusing groups: pregnant women's geographical experiences of Hamilton, New Zealand/Aotearoa. *Area* 28(2), 143–9.

McDowell, L. 1992: Valid games? A response to Erica Schoenberger. *Professional Geographer* 44, 212–15.

Miles, M. B. and Huberman, A. M. 1994: *Qualitative data analysis: an expanded sourcebook*. London: Sage.

Routledge, P. 1997: The imagineering of resistance: Pollock Free State and the practice of postmodern politics. *Transactions of the Institute of British Geographers* NS 22(3), 359–76.

Schoenberger, E. 1991: The corporate interview as a research method in economic geography. *Professional Geographer* 43, 180–9.

Sporton, D. 1999: Mixing methods of fertility research. *Professional Geographer* 51, 68–76.

Valentine, G. 1989: The geography of women's fear. *Area* 21, 385–90.

—— 1992: Images of danger: women's sources of information about the spatial distribution of male violence. *Area* 24, 22–9.

—— 1999a: Being seen and heard? The ethical complexities of working with children and young people at home and at school. *Ethics, Place and Environment* 2, 141–55.

—— 1999b: Doing household research: interviewing couples together and apart. *Area* 31(1), 67–74.

WGSG (Women and Geography Study Group) 1997: *Feminist geographies: explorations in diversity and difference*. London: Longman.

# 4

# Negotiating the personal and the political in critical qualitative research

## Audrey Kobayashi

Critical scholarship often involves reflexivity on the part of the researcher, who is strongly aware of her own role in constructing the social systems that we study. In this chapter I reflect on some of the reflexive challenges I have faced, and on some ways of using qualitative methods to meet those challenges. Through relating a series of actual research experiences, I try to convey a sense of just how complex are the ethical, ideological and methodological issues faced by the geographer in the field.

I define 'critical' scholarship as scholarship that conveys the social consequences of the situations that we study, and that attempts to uncover the tensions and contradictions faced by people in those situations. Critical scholarship takes a position on what is, and on what could be, as well as providing a theoretical understanding of the systemic ways in which social relations are constructed. I also take the ethical position, however, that it is necessary (for me, at least) to go beyond critical scholarship, which explains contradictions in the world, to *activist* scholarship, which attempts to resolve those contradictions, to bring about actual change. There is a wide variety of ways of doing activist research, but activist scholarship usually involves working closely with community groups. It can be messy and unpredictable, and can present deep ethical challenges. It requires reflexivity on the part of a researcher, who must examine her own motives, and the effects of her actions, as both researcher and activist.

Once we assume a reflexive position, the ethical questions that follow are not straightforward. They may even be contradictory. They may often be at odds with what is conventionally defined as ethical in social-science research. A basic assumption of ethical guidelines as they exist in most universities, for example, is that the researcher should attempt to disrupt the lives of her subjects as little as possible. Furthermore, an established assumption of good methodology conventionally requires that the researcher maintain a neutral position with respect to subjects so as not to 'taint' the data obtained in the course of research. This chapter begins with a critical look at these two assumptions.

Developments in scholarship during the 1990s posed a potential challenge to these two established principles. Although the question of 'relevance' and social justice has been a significant thread in human geography since 1970,[1] Iain Hay, in a paper in the inaugural issue of the journal *Ethics, Place and Environment*, exhorts geographers to give greater consideration to ethical concerns in research. But to do so, he suggests, means not to accept inflexible codes of ethics but, rather, to possess a 'moral imagination', based on 'sets of prompts intended to encourage informed thought about ethical practice' (Hay 1998: 56). Such prompts would allow a flexible approach to ethics, founded in moral contemplation rather than in rigid codes. To do so, however, raises serious questions concerning the socially agreed-upon principles that guide such codes. Where moral commitments differ, whose are more important, those of the researcher or those of the subject? Why is privacy held in most ethical codes to be a principle that stands above others as inviolable, and how do we deal with cultural variations in standards of privacy? What do we do in situations where we find our own moral codes, or those of the institutions that employ us, compromised?

Another potential contradiction is presented by funding agencies, which place more and more emphasis on policy relevance and community participation in research projects. I recently became involved as principal investigator in a project funded by the Social Sciences and Humanities Research Council of Canada under a programme entitled *Social Cohesion in an Age of Globalization*.[2] The criteria for funding included policy relevance, and having strong partnerships with community organizations. The very title of the programme, *Social Cohesion*, of course implies a normative agenda. My personal approach to research supports these criteria. Yet, in the course of developing the project many ethical questions arose. Whose public policy agenda are we serving? Need the agenda of the researchers fit with those of the government of the day? What forms of relationship with communities are appropriate? Can we work in full partnership with community groups and, if so, how do we negotiate the policy approaches and the desired outcomes among three sets of interested parties, the government, the community and the researcher?

The most difficult contradiction, however, occurs for those academics who view their work as activist. A *critical* perspective transcends methodology to view qualitative methods as a basis for challenging dominant ways of understanding, and for exploring the contradictions that give rise to social inequities and patterns of marginalization. It demands an ethical positionality. I wish to go a further step, however, to suggest that *activist* scholarship requires a commitment on the part of the researcher to become involved in the goal of social change. Such scholarship makes no claim to neutrality or non-intervention. It is by definition normative and undertaken with the hope that the research will result in changed conditions for the subjects – not just the world-in-general, but those particular subjects with whom one works on any specific research project. Many university ethics review committees have difficulty with such an avowed objective, because it directly contravenes claims towards neutrality or non-intervention on the part of the researcher. This does not mean, however, that activist-oriented research lacks ethics; quite the opposite.

The ethical questions involved in overtly activist research are many. How does one negotiate between a specific social objective and the need for verifiable knowledge? How does one deal with the inevitable emotional aspects of working with people whose circumstances may require immediate attention? How does one juggle time commitments between community activism and the more solitary pursuits of research and writing? What does one do when one's objectives as a researcher, or as an ethical individual, do not accord with those of one's subjects? How does one overcome the inequality between researcher and subject that is imposed by the researcher's position of relative power, which may be expressed in terms of class, education or access to resources?

I cannot address all of these questions here, but this chapter outlines some of my personal research experiences in going beyond critical, to what I term *committed* scholarship. Qualitative methods, a critical perspective, and activism for social change are mutually interdependent aspects of committed scholarship. I shall also argue in this chapter that there is an ethics of commitment.

## The need for critical qualitative methods

For some time now, social scientists have recognized a need for qualitative analysis, but most textbooks on the subject justify this need in terms of the kinds of data needed for more in-depth research than can be obtained from quantitative data. Some of the leading textbooks on the subject illustrate the problems of undertaking qualitative methods without taking a critical perspective. Patton (1987: 48) stresses that 'central themes in qualitative methods are the emphasis on depth and detail', and goes on to enumerate the value of data gained through in-depth interviewing, focus groups and field observation. Of the 'personal' side of qualitative methods, he observes:

> Interviewing people can be invigorating and stimulating. It is a chance for a short period of time to peer into another person's world. A good interview lays open thoughts, feelings, knowledge, and experiences not only to the interviewer but also to the person answering the questions. The process of being taken through a directed, reflective process affects the person being interviewed. It is not unusual for an interviewee to say, 'you know, I hadn't thought of that for a long time'. As respondents think about questions they may surprise themselves with fresh insights, previously unarticulated concerns, and new ideas.
>
> (Patton 1987: 140)

Patton's account seems highly insensitive, not to say ethnically questionable. He provides little reflection on what impact these in-depth interviews may have had on the participants. Two things about Patton's approach are deeply troubling. The book says absolutely nothing about the ethics of peering into other people's lives and 'laying bare' their emotions, of the consequences, ethics and responsibilities of the researcher who may be changing the lives of her subjects in significant ways. Nor does it say anything of the researcher, whose life may also be changed. Qualitative methods are presented as nothing more than a means of collecting better data, in order to achieve an unspecified and uncritical 'truth'.

Another text, by Miles and Huberman (1994), provides an extensive chapter on ethical issues of qualitative research, including a review of writing on the topic (see especially House 1990; Flinders 1992; Sieber 1992) organized under the following categories: utilitarian (informed consent, avoidance of harm, confidentiality); deontological (reciprocity, avoidance of wrong, fairness); relational (collaboration, avoidance of imposition, confirmation); ecological (cultural sensitivity, avoidance of detachment, responsive communication). Again, these are important issues, over which most qualitative researchers have spent a great deal of time pondering. But the fact that they are presented in a final chapter sets them outside the research itself, as *principles* that guide research actions, rather than as the ontological starting point. The excellent discussion of how to do ethical research, moreover, is concerned almost entirely with actions of the researcher towards the researched, rather than with questions of how the research connects with the life of the researcher.

Other discussions of qualitative methods, although they do not address the questions of how qualitative research draws the researcher into the web of commitment, are explicit about the need to consider the needs of research subjects as more than information vessels. Noblit and Engel (1992) emphasize the 'moral imperative' to be holistic in qualitative ethnographic research, based on Marcus and Fischer's pathbreaking admonition (1986) to contextualize research. By 'moral imperative', they mean the moral obligation to recognize that subjects' lives are multifaceted, interconnected, contextually situated and deeply meaningful, in ways that cannot be conveyed easily by simple descriptions such as those achieved quantitatively. Following the moral imperative both provides a systematic sense of the connections between individual experience and the social whole (a critical perspective), and provides a deeper understanding of the meaning systems in which those experiences are embedded. Such an approach requires sensitivity to the ways in which different subjects construct meaning, and recognizes that the researcher needs to be aware how emotional responses may occur (Sieber 1993). Such issues have received attention for specific groups, including the elderly (Golander 1992), children (Herzberger 1993), those who have experienced trauma (Curran and Cook 1993; Kennedy Bergen 1993), marginalized groups such as gay men with HIV-AIDs (Martin and Dean 1993) or racialized communities (Henderson 1994 ). All these studies show not only the need for sensitivity to subjects' needs, but also that those needs vary according to time, place and group.

In contrast, overtly feminist discussions go beyond contextual understanding, to be more explicit about the potentially *transformative* nature of qualitative research. Most feminist researchers not only recognize holistic lives, but pay especial attention to questions of difference and reciprocity, and to the importance of relational and situated knowledge. Although these considerations do not, in themselves, challenge the ontological status of the relationship, they extend and amplify the list of ethical concerns enumerated by Miles and Huberman above. For example, Anderson et al. (1990) use Dorothy Smith's concept (1975) of 'beginning where we are' to stress, inter alia, our obligation to help women tell their stories for the benefit of all women, using their stories to learn to hear differently and to change our own lives. Such stories inform feminist analysis in (at least) two ways, by providing a means for women to change their lives in speaking their experiences, and by providing a basis for critical feminist scholars to become involved in that change. Such scholars recognize that doing feminist research means taking a political stand, in the sense not only of being wary of entanglements in unexpected political situations (cf. Sieber 1993), but also of the all-encompassing nature of research as activism. Research, for feminists, is meant to engage and to change the world, not simply to describe it with depth and sensitivity. No matter how sensitive our methods, moreover, our research produces direct changes in the lives of our subjects. The extent of those changes, and the force of their impact, depend strongly upon ethical as well as methodological choices that guide every exercise in qualitative methods.

## Getting started: what is ethical about qualitative methods?

In 1981 I was in a Japanese village, going from house to house seeking information about the history of emigration to Canada. Trained as a historical geographer, I was keenly determined to collect accurate data and aware of the value of first-hand accounts. I approached a very elderly woman, bent from years of working in the rice fields, her hands twisted and gnarled with arthritis. She eyed me suspiciously, at first unwilling to speak to me. Knowing that there had been several emigrants from her household, however, and that her age should place her among the generation of emigrants, I cajoled and pleaded, told her how important it was for me to obtain this information, and that the study was supported by a prestigious Japanese university.

Most of the people I had interviewed up to that point had been either men who were quite proud of their accomplishments in travelling to Canada during their young years as migrant workers, or men and women whose parents and grandparents has gone to Canada, creating the conditions for a more comfortable life back in Japan. They had on the whole been quite open, willing to talk about a range of experiences, balancing the positive and the negative with a sheen of nostalgic pride. I was excited that for the first time I had the opportunity to interview a woman who might have been part of the migrant experience.[3]

In my excitement to peer into her world, however, the more I cajoled, the more reticent she became. As I was about to give up, she suddenly burst out: 'Why are you coming here talking to me about that!? It was horrible! I was treated like an animal by those white people! I'm not an animal! Get out of here!'[4] Chastened and shaken, I retreated as quickly as possible. In my subsequent rounds of the village houses, I avoided her home. When I had to walk by, I checked unobtrusively first, to make sure that she was not working out in her garden.

### Lessons learned

I was deeply disturbed by this encounter, but it was some time before I realized the fuller implications of the exchange. The experience reinforced the fact that important lessons are learned in the field that could never be conveyed through reading. It jolted me out of the complacent belief I had developed as a graduate student that I could control the information that would comprise the data of my dissertation. The incident also nudged me into feminist scholarship (scarcely recognized by geographers in 1981), and made me aware of the profoundly different ways in which men and women experience similar external conditions. It also made me realize not only that the experience of racism is profoundly hurtful, but that racism is structured along lines of class and gender, a realization that was to become a foundation of my subsequent research career.

Finally, the incident taught me in addition that research ethics are not just about following the rules. They involve a need to care deeply about the people who become participants in a research project, even if it is only to the extent of refusing to participate. They involve an ethical commitment, not only in the larger sense of making the world a better place but in the more immediate sense of understanding and taking responsibility for how one sets in motion the complex emotions that flow back and forth in the course of a research encounter, sometimes reaching beyond that moment to affect people's lives in more permanent ways, over which we may have no control. There was nothing that I could do to change the experiences that the woman had had in the past, but had I had a different kind of training or been better prepared to undertake such fieldwork, I might have been able to respond better to what was clearly an expression of pain nearly fifty years after the fact. While I would not have described myself as an activist researcher at that time in my career, therefore, this incident was a very important one along the road to becoming an activist.

The ethical guidelines set out by most universities do not address the situation I faced on that occasion. The guidelines of the University of California, where I did my doctoral research, did contain a vaguely specified reference to the need not to cause 'harm' to one's subjects, but that sense of harm was developed to refer to adverse effects that might result through research based on medical procedures that might be detrimental to the well-being of the subject. Sadly and ironically, harm clauses in ethical guidelines are as much (or perhaps more) concerned to protect the interests of the research institution against legal suits as they are to protect the subject from emotional ill effects. This issue in itself is an ethical one. The outcome would have been much better had those guidelines been less concerned with protecting university interests, and more concerned to develop in the researcher a sense of caring and commitment. They might also have included some advice on how to deal with unexpected field situations.

**Fig. 4.1** *Combining research and community activism: Audrey Kobayashi at a meeting of the National Association of Japanese Canadians, October 1999*

## Mixing messages: the politics of politics

The example I present above came about in the course of a fairly conventional research project, in which I was at that time ill prepared for unexpected ethical challenges. Some years later, I entered a more explicitly political situation, albeit also one that was unexpected. I was approached by the National Association of Japanese Canadians (see Fig. 4.1) to assist them in providing background research in support of their attempt to obtain redress from the Canadian government for actions taken against Japanese Canadians during the 1940s.[5] I agreed, thinking at first that my research on the demographic effects of the uprooting and internment of Canadians of Japanese ancestry would provide irrefutable and unbiased evidence upon which the negotiations could proceed.

I was still not working as a committed activist. Within a matter of months, however, I found myself at the negotiating table, using my careful research as the basis for arguments set before the government in stating our case. I was one of a group of about six individuals, fiercely determined not only to achieve a fair deal for about 17,000 individual survivors of the government's actions, but also to advance the cause of anti-racism in Canadian society. Our arguments were framed in terms of social justice, healing and the need to achieve lasting effects in overcoming racial discrimination.

Contradictions? Not really, given that we were all on the same side, believed firmly in our position and were convinced of the moral rightness of our cause. But clearly I was no longer involved in anything approaching 'neutral' research. My data would not sit gathering dust on a shelf, or be consigned to the pages of academic journals. It was immediately thrust into the political forum. Negotiators on behalf of the government also believed in the justness of the cause, although they held a firm line in terms of the amount of compensation that they were prepared to provide. But this experience shattered forever any belief I might have held that research can be value-neutral or disengaged from social change. Certainly, this project was more deeply engaged with the politics of change than most, and so the ideological implications were more obvious, but it nonetheless reinforces a number of general principles about the ethical relationship between the personal and the political in a research context.

One of the most difficult barriers to full participatory action on the part of activist researchers is the inside/outside division. In this case, I found myself on the inside, in the sense that I was speaking for a community of which I was a part.[6] That community, however, like all communities, was diverse, and covered a diverse political spectrum. While we had strong support for our negotiation on behalf of a group consisting of more than 17,000 people, not all of them supported us. We had as much work to do within the community as outside, therefore, labouring always with the knowledge that people possessed different levels of knowledge, different experiences, and different reasons for supporting or not supporting the cause. Class, gender and a range of personal factors influenced this diversity. We agonized over issues of representation and legitimacy, as well as power relations between the negotiating committee and the community in general.

This situation left us with a difficult challenge with respect to our government counterparts. We spoke *for* the community, but not with a single voice. We had, therefore, simultaneously to represent the diversity of the community and to present its common concerns, while resisting attempts on the part of government officials to essentialize Japanese Canadians as a singular social entity. It was tempting to resort to 'strategic essentialism' – that is, the presentation of the group *as though* their shared cultural background imposed immutable and inevitable social characteristics – but to do so would have denied not only the community's diversity but also our fundamental anti-racist position that shared ancestry does not result in essential features.

Perhaps I should make a brief intervention here about the issue of essentialism. Anti-essentialism, or the belief that human characteristics are not determined by so-called biological conditions such as 'sex' and 'race', is a cornerstone of both feminist and anti-racist research. It is both a theoretical position, in that it theorizes the nature of human being as irreducible to 'essences' that create necessary differences among human beings – for example, making women 'weak' or people of colour 'uncivilized'. It is also a moral position in that it provides a position from which to fight the effects of inequality based on sexism or racism, both of which are forms of essentialism. But, once we give up all essentialist views, it can become very difficult to find a space from which to argue for the specific needs of oppressed groups.

We overcame this paradox by taking a two-pronged approach. On the one hand, we insisted that each individual affected by the settlement be treated as an individual Canadian citizen. Based on evidence derived from demographic material, focus groups held in the communities and extensive archival material, we argued that the government's actions in uprooting and interning more than 20,000 Canadians of Japanese ancestry had been based on a racist attempt to create an all-white society, and that that attempt violated the human rights of those individuals, as they were defined both in the 1940s and in the 1980s. This individual model was based on a particular definition of human rights as being held by individual human beings, while at the same time it thwarted essentializing strategies on the part of government negotiators.

On the other hand, however, we needed a geographical solution to the issue of community. We argued that, while people exercise membership in communities as individuals, the concept of community, of sharing based on an agreed-upon cultural commonality, is a legitimate one. While again not wishing to essentialize community as anything but the result of collective actions, we argued that communities represent specific places, organized to have meaning for significant sociocultural groupings. Based on supporting evidence that the government had undertaken a deliberate and systemic programme of destroying community institutions, and arguing that such institutions are necessary if individuals are to participate effectively as community members, we then negotiated an additional settlement on behalf of the community, in order to allow for the rebuilding of institutions and infrastructure.[7] By thus adopting a theoretical understanding of the contextual, and by acting upon this theoretical position, we avoided having to resort to 'strategic essentialism', or behaving as though it were essential features, rather than their common experiences, that bound Japanese Canadians together.

### Lessons learned

This example falls outside the purview of standard academic research projects, but, for me, it represented a turning point in the development of my commitment to an activist research agenda. I learned that no research project is without a political purpose, although some may be more forceful or more obvious than others. I learned that 'data' can take many forms and are subject to a wide variety of readings, and that the researcher need not necessarily exempt herself from influencing the ways that others read her work. Most of all, I learned that theory and practice can be very closely related, can inform one another and, indeed, must do so in the interests of both good scholarship and social justice. But, if such goals are to be achieved, it will not be through short-term commitments, and one-off research projects. The combination of research and activism requires a commitment for the long haul, the need to immerse oneself in the community situation and more connection to the community than most researchers have. Such research is best carried out, therefore, as in my case, close to home, in the surroundings in which one's voice carries meaning because it is part of the place wherein its speaks.

## Getting political without getting personal

My third example takes a step back from community entanglements. My colleague and I were involved between 1997 and 1999 in a project to compare employment equity policies in Canada's ten provinces.[8] In the course of conducting fieldwork that involved interviews with employment equity experts employed in the provincial public services, we encountered several ethical issues.

The interviews were unstructured, but followed a set format, asking a number of questions about best practices for achieving equity in the workplace. We had had two expectations of these interviews: that they would indicate to us some of the regional variations in political and bureaucratic culture that explain differences in approaches to employment equity, and that the experts would provide guidance on how employment equity policies might be better implemented. Following the ethical guidelines of Queen's University, we had presented waivers to the participants, with our written assurance that we would not identify them in our final report without their permission.

We were correct in our first hypothesis. In the second, however, we were only partially correct. Some of the experts were indeed deeply committed. Others were not only minimally committed, but in fact expressed views that we considered racist and sexist. These findings were disturbing, and presented us with some significant research dilemmas. How could we proceed to interview people whose opinions were so different from our own, and how could we then write up the results? Should we engage with the people we interviewed, argue with them, try to change their opinions? Should we change our own opinions in response to the new information we were receiving?

The first decision we took was to remain as neutral as possible during the interviews themselves – that is, to ask questions in such a way that we did not lead the interviewee, and to refrain from giving personal opinions on the issues under discussion. Neutrality is not, however, possible. One can never escape from – try as one may to mitigate them – the emotive qualities that characterize any human encounter, including the interview. Simply put, no matter how we tried to standardize our approach, we had more rapport with some than with others. Some were more talkative or experienced than others, some more withdrawn, some more committed, a few hostile to our stated intentions. We asked the same questions of everyone, however, kept the timing of the interviews as uniform as possible, and tried to develop a similar attitude, while at the same time recognizing that variations in human demeanour, as well as in ideology, represent a significant factor in our analysis of how to achieve equity in the workplace.

We faced another challenge when it came to interpreting the data. We had decided early that we would refrain from identifying any of our respondents, although some of them clearly would have liked to have been named, and although it became very difficult to write about particular provincial programmes without naming those who had developed them. We had somehow to write about the racism we had encountered, without violating our own ethical standards but in a way that would convey how problematic these attitudes were for achieving equity.

Psychologists have attempted large-scale analyses of racist attitudes with some success. Berry and Kalin (1997), funded by the Canadian government, undertook thousands of telephone interviews in an attempt to analyse racist attitudes and to develop a scale of tolerance. These interviews were undertaken in anonymous conditions, based on a limited questionnaire. The results are significant for understanding the phenomenon of racism broadly, but a deeper understanding of racism would require extensive follow-up, on a smaller scale and using qualitative methods. Such qualitative methods could not avoid the issues outlined above – the fact that racism is a complex and disturbing psychological condition, that conversations about racism, no matter what controls are imposed by the researcher, can be uncomfortable, emotional and at times even dangerous. In the course of such conversations, the researcher is bound to have radically different opinions from some of the subjects. She may even have to confront her own dislike of them. At the same time, if the research is to have any worth and meaning, a relationship must be maintained at least for as long as is necessary to complete the interview, and the researcher must have the freedom to report on the results, in an ethical manner that maintains confidentiality but that does not thereby compromise the conclusions.

Well, our conclusions were compromised nonetheless, at least to the extent that they were *affected* by our need to protect our subjects. We decided that, where the names of individuals might be deduced from the context (despite our efforts at anonymity), it would be only in the more positive examples. This did not mean, however, that we would report only the positive aspects of the work, for we felt that we needed to discuss the various ways in which the 'experts' can present a barrier to equity. We did so by paraphrasing their opinions, and by disassociating them from particular places. In other words, where we encountered a particularly disturbing set of comments, we did not reveal from which provincial setting the person or persons expressing those comments originated. The result of this decision was that we lost an opportunity to discuss the geography of equity, by *placing* comments within a particular sociocultural context. Our analysis, at least in this section of the report, was place-less, but we felt an ethical imperative to make this sacrifice.

### Lessons learned

This research project illustrated for us the complexities of working in an area that is politically and ideologically fraught. Unlike the example with Japanese Canadians, where I could assume at least a high level of common experience around which to mobilize community needs, this situation was characterized by the diversity of opinions, experiences and political contexts that we encountered at different field sites. This was a strikingly geographical example of how local context produces different experiences, and of how common experiences shape locality.

This example also highlights some key points concerning qualitative research. It is possible to do research with individuals across a wide ideological spectrum, but it is important to be as consistent as possible in addressing those with whom one does not have a high level of personal rapport, and those with whom one establishes rapport. Doing activist research does not mean being explicitly activist at every stage. Here, we hoped that our work would help to change the lives of the workers represented at the various sites, but we did not feel that the way to do so was to engage in direct confrontation with those we interviewed. Finally, social processes such as racism and sexism are very complex, and sensitive interviewing techniques are required to begin to probe the intentions, experiences, beliefs, yearnings and contextual circumstances involved in the creation of raced and gendered human beings.

## Conclusion: learning from the lessons learned

These examples provide a range of lessons for qualitative research. The first example shows that qualitative work inevitably involves interpersonal relationships between the researcher and the subject that may be intensely emotional, although not always in a positive manner. Such experiences may affect the interview subject more strongly than the researcher, by bringing up old memories, in this case. There is an ethical obligation, of course, to avoid such emotional harm where possible, but, where it is not possible (because, in this case, for example, it was unexpected), there is also an obligation to be prepared to follow through in a manner that is appropriate to the situation.

In the second case, among the many lessons learned is that qualitative methods combined with activism may also have unexpected results, but will inevitably be complicated by the political relations among a range of subjects, including individuals, communities and authorities. While we may feel an ethical and a scholarly commitment to discover 'truth', that truth is highly variable, looks different depending upon individual perspectives and is always subject to ideological manipulation. This example also illustrates the need to avoid essentializing human beings as a matter not only of social theory, but also of social justice.

The third example illustrates the difficulties of conducting qualitative research among subjects whose perspectives may vary considerably from that of the researcher. There is an ethical commitment to be fair to such subjects, even when the researcher may take a strong dislike to them. But there is also a need to present the results of such work in such a way as to encourage deeper understanding of the more unpleasant aspects of human behaviour. Such understanding is a necessary component of understanding the larger picture in which racist (or other oppressive) relations are structured.

All three examples have led me to a stronger commitment towards research as activism. Activist research means two things for me. First, it involves a critical perspective on the world, through an attempt to understand the systemic ways in which human relations result in oppression, and especially in racialized oppression. It also involves, however, going beyond a critical perspective to make my research a very personal undertaking. It is personal because it involves the range of emotional experiences involved in qualitative research, because it requires me to take a personal stand on contentious ideological issues, and because it involves a personal confrontation with the very racist conditions that I hope to study. It also involves a deep personal commitment to make a difference, to change the world in some small way, if my research is to have meaning.

## Addendum: essentialism

The most difficult theoretical question I have faced concerns essentialism. A social constructionist position posits that nothing is essential – that is, nothing is absolute or determined. Everything that has any meaning for human beings (that is, everything that has come under human perception or cognition) is socially constructed, to the extent that it is brought into the realm of discourse. Discourse is the social act of creating meaning, simultaneously transforming (however minutely) the world and ourselves, and infusing the relationship with power. That power is almost always to some degree differential. Of course, it is a very different thing socially to construct a rock than it is socially to construct another person.[9] To engage another person, on the other hand, is to engage in a mutually constitutive exercise, by which we define one another, albeit under conditions of inequality or, in some cases, dominance or oppression.

There are no points of overlap between essentialism and social constructivism. Either, as I believe, everything is part of the process of social transformation, at whatever insignificant or even undetectable scale, or everything is ultimately ordered by some form of external, ineffable and controlling truth. Some authors have attempted to make a distinction between those things that are essential, or necessary, and those that are socially constructed, or contingent. This approach is fundamentally essentialist. Not only is such an additive model empirically unverifiable, but it also trivializes the actions of human beings, relegating the act of social construction to some kind of mental sphere, where people imagine themselves out of conditions that are basically determined. Social constructionism, in other words, is not an idealist theoretical position but, rather, one that conforms to a materialist interpretation of the world. A materialist interpretation recognizes all human acts, as well as all objects (including bodies) as materially, but socially, produced.

To assert the universal quality of social construction, however (and this is about the only context in which I would use the term 'universal'), is not to give social construction (or, to put it another way, human agency) the power to effect outcomes. Human actions are always mediated by the actions of others. They are historically cumulative. Our actions escape our power, to varying degrees, as soon as they are acted. One of the most important goals of social science, therefore, is to understand the conditions under which human actions have more or less efficacy, more or less power to affect others and to effect change.

Overcoming essentialism has been of overwhelming importance both for feminism and for anti-racism for more than theoretical reasons. As critical scholars committed to bringing about social change, we can achieve change only by understanding what is possible. Such understanding requires not only a solid theoretical sense of how we believe social relations to occur, and how we understand human being to be constituted. It also requires a powerful vision of change, a commitment to make things happen and a sense of a different world. Such a vision can never be essentialized.

## Notes

1. For a review of the early 'relevance' debate, see Kobayashi and Mackenzie (1989).
2. I wish to acknowledge the support of the Social Sciences and Humanities Research Council of Canada, Strategic Grant 929 99 1012, 'Transnational Citizenship and Social Cohesion: Recent Immigrants from Hong Kong to Canada'.
3. Approximately two out of three households in this village had sent migrants to Canada during the early years of the twentieth century. Among the first generation of emigrants, women migrants in particular were strongly differentiated by class. In the wealthier households migration was considered unfitting for women, except for the wives of non-inheritor sons. Among the poorer households, women migrants could contribute to household income by working as domestic servants for white families in Vancouver.
4. This is a paraphrasing of the original discussion, in Japanese.
5. For more complete accounts of this process, see Kobayashi (1987, 1992).
6. I did not, however, stand to benefit directly from the redress settlement, since I had been born just after the period in question.
7. The final settlement included an acknowledgement of wrong, a symbolic payment of $CAN21,000 to each individual affected by the government's actions, plus a community settlement of $CAN12 million, to be administered by a foundation set up by the National Association of Japanese Canadians, and distributed to community groups, as well as an endowment for a Canadian Race Relations Foundation, intended to fight racism in the future.
8. What is officially referred to in Canada as employment equity, is called affirmative action in the USA and equal opportunities in the UK. The study is reported in full in Bakan and Kobayashi (1999). I wish to acknowledge the significant contribution made by my research partner, Abigail Bakan, to this section of the chapter. It is another aspect of qualitative research that it is often enhanced through collaborative work.
9. Rocks are nonetheless transformed by human constructions. Consider, for example, the Rock of Gibraltar, which has been given such an important role in the constitution of world relations of power. Consider also the diamond, a 'rock' of considerable power, symbolism and emotive quality.

## Key references

Katz, C. 1992: All the world is staged: intellectuals and the project of ethnography. *Environment and Planning D: Society and Space* 10, 495–510.

Kobayashi, A. 1994: Coloring the field: gender, 'race', and the politics of fieldwork. *Professional Geographer* 45(1), 73–80.

# References

Anderson, K., Armitage, S., Jack, D. and Wittner, J. 1990: Beginning where we are: feminist methodology in oral history. In Nielsen, J.M. (ed.), *Feminist research methods: exemplary readings in the social sciences*. Boulder, CO: Westview, 94–112.

Bakan, A. and Kobayashi, A. 1999: *Employment equity policy in Canada: an interprovincial comparison*. Ottawa: Status of Women Canada.

Berry, J. and Kalin, R. 1997: Racism in Canada: evidence from national surveys. In Driedger, L. and Halli, S. (eds), *Race and racism in Canada*. Ottawa: Carleton University Press, 1–16.

Curran, D. J. and Cook, S. 1993: Doing research in post-Tiananmen China. In Renzetti, C. M. and Lee, R. M. (eds), *Researching sensitive topics*. Newbury Park, CA: Sage, 71–81.

Flinders, D. J. 1992: In search of ethical guidelines: constructing a basis for dialogue. *Qualitative Studies in Education* 5(2), 101–16.

Golander, H. 1992: Under the guise of passivity: we can communicate warmth, caring and respect. In Morse, J. M. (ed.), *Qualitative health research*. Newbury Park, CA: Sage, 192–201.

Hay, J. 1998: Making moral imaginations: research ethics, pedagogy and professional human geography. *Ethics, Place and Environment* 1(1), 77–92.

Henderson, J. N. 1994: Ethnic and racial issues. In Gubrium, J. F. and Sankar, A. (eds), *Qualitative methods in aging research*. Thousand Oaks, CA: Sage, 33–50.

Herzberger, S. D. 1993: The cyclical pattern of child abuse: a study of research methodology. In Renzetti, C. M. and Lee, R. M. (eds), *Researching sensitive topics*. Newbury Park, CA: Sage, 33–51.

House, E. R. 1990: The ethics of qualitative field studies. In Guba, E. G. (ed.), *The paradigm dialog*. Newbury Park, CA: Sage, 158–64.

Kennedy Bergen, R. 1993: Interviewing survivors of marital rape: doing feminist research on sensitive topics. In Renzetti, C. M. and Lee, R. M. (eds), *Researching sensitive topics*. Newbury Park, CA: Sage, 81–196.

Kobayashi, A. 1987: Real or apprehended? The Japanese–Canadian redress issue and human rights. *Human Rights Advocate* 3(10), 1–3.

—— 1992: The Japanese–Canadian Redress Settlement and its implications for 'race relations'. *Canadian Ethnic Studies* 24(1), 1–19.

—— and Mackenzie, S. 1989: Introduction: humanism and historical materialism in contemporary geography. In Kobayashi, A. and Mackenzie, S. (eds), *Remaking human geography*. London: Unwin Hyman, 1–16.

Marcus, G. E. and Fischer, M. M. J. (eds) 1986: *Anthropology as cultural critique: an experimental moment in the human sciences*. Chicago: University of Chicago Press.

Martin, J. L. and Dean, L. 1993: Developing a community sample of gay men for an epidemiological study of AIDS. In Renzetti, C. M. and Lee, R. M. (eds), *Researching sensitive topics*. Newbury Park, CA: Sage, 82–100.

Miles, M. B. and Huberman, A. M. 1994: *Qualitative data analysis: an expanded sourcebook.* London: Sage.

Noblit, G. W. and Engel, J. D. 1992: The holistic injunction: an ideal and a moral imperative for qualitative research. In Morse, J. M. (ed.), *Qualitative health research.* Newbury Park, CA: Sage, 43–9.

Patton, M. Q. 1987: *How to use qualitative methods in evaluation.* Newbury Park, CA: Sage.

Sieber, J. E. 1992: *Planning ethically responsible research: a guide for students and internal review boards.* Applied Social Research Methods Series, 31. Newbury Park, CA: Sage.

—— 1993: The ethics and politics of sensitive research. In Renzetti, C. M. and Lee, R. M. (eds), *Researching sensitive topics.* Newbury Park, CA: Sage, 14–26.

Smith, D. 1975: Women and psychiatry. In Smith, D. and David, S. (eds), *Women look at psychiatry: I'm not mad, I'm angry.* Vancouver: Press Gang, 1–20.

# PART 2

## Interviewing

PART 2

# 5

# Shared lives: interviewing couples, playing with their children

## Stuart Aitken

## Introduction

In-depth interviewing is attracting considerable attention in geography, and, as our knowledge of the method broadens, our questions and concerns reflect greater sensitivity to a complex set of personal, political and place-based processes (Katz 1994; Valentine 1997). For example, geographers are raising specific questions relating to where interviews take place (Elwood and Martin 2000), body metaphors and gender relations (Sparke 1996) and the interpersonal communicative dynamics involved in separate conversations or group discussions (Goss 1996; Longhurst 1996). Some of this literature is still a bit mechanistic (the 'how-to' variety), but for the most part geographers are sensitive to the quirky and complex politics that surround sitting down with people (who are usually strangers) and talking about their lives. In this chapter I wish to question some of the ways interdependencies between interviewers and cohabiting partners inform research and practice. To do so, I draw from a study of families in San Diego that focuses upon the gender and place-based changes stimulated by the birth of a child.[1] The issues in this chapter are raised primarily from my own experiences and from those of research assistants who interviewed household members along with me in what became known as the *Family Fantasies* project. To illustrate some points I use quotes from interviews, personal communications with student interviewers and excerpts from field logs.[2] In places I touch upon various methodological logistics, but that topic is covered well by Gill Valentine (1999), who concludes that no one strategy for interviewing partners is superior to another. Rather, I highlight the political and ethical messiness of encountering shared lives through interviews that are periodic, intense and intrusive. Often this messiness takes the research on unexpected turns that become the most interesting and provocative aspects of the work.

To organize the chapter I begin by briefly discussing what we know about interviewing couples and where geography contributes to this literature. I then outline the genesis of the *Family Fantasies* project, with an emphasis on being in the 'field' with cohabiting partners. Issues of motivations and expectations are outlined in this preliminary discussion. The body of the chapter focuses on different aspects of communication as they arise in an interview setting. The chapter ends with a brief capstone comment on encountering difference as a moral imperative.

## Do you really want to get that involved?

Grappling with how to interview couples and household members has a long history in family studies. A protracted debate pivots around the various methodological pitfalls and benefits of interviewing household members separately or together (cf. Allan 1980; Hertz 1995; Handel 1996). This debate, like others outside geography, seems to focus on the mechanics and logistics of interviewing rather than how the interview reflects and refracts the nuances of people's shared lives and their jointly negotiated stories. I agree with Valentine that this oversight results in a lack of 'attention both to the power-laden and ethical consequences of probing joint stories, and to exploring the complexities and contradictions of the contested realities of shared lives' (1999: 73). When our attention turns to households and families, we need to be particularly sensitive to power relations between household members and how our presence (and that of our study) is contextualized within those relations. Focusing on shared lives and attempting to unravel family members' stories that are joined spatially and experientially raise serious questions. For the most part, people who choose to share a domestic space consciously and unconsciously negotiate activities and events. There is tension, compromise and equanimity in cohabiting space that partners usually articulate with difficulty. The kinds of questions that help unravel this complexity depend upon your project's focus, but it is useful to think about how your approach to the interview context changes the parameters of partners' relations. I begin, then, with a focus on initial motivations and the potential intrusiveness of research questions and then move to a discussion of the politics of the interview context itself.

## Getting started

In qualitative research with partners, you position yourself at the margins of an unfathomable set of interpersonal politics from where you scratch only a small fragment, which you hope is sufficient to answer your research questions without damaging the relationships you want to understand. Interviews with partners are potentially invasive and so a good place to begin is with your own motivations and expectations.

### Research motivations and expectations

Your motivations may cover the gamut from procuring funding to fulfilling degree requirements to crusading for a better world, but it is important to recognize how these motivations are transformed into expectations, and how expectations become embedded in interview contexts. One of my initial motivations for the *Family Fantasies* project was methodologically driven. I expected that the birth of a new child would heighten a caregiver's acuity about his or her place in the world and provide an exceptional context for discussing families, communities and place-based politics. But I had other motivations. It was no coincidence that I had just become a father myself and was interested in learning how other parents coped with the love, hate, frustration, hope, fears, denial and anxiety that I was experiencing. It turned out that, during the project, a growing honesty about this motivation helped me to connect more fully with some interviewees.

I was also motivated by what I thought was a fundable set of ideas. The questions I posed seemed relatively innocuous but nonetheless interesting to the United States' National Science Foundation (NSF) and my university's review board on human subjects. The project secured approval and funding, but the expectation of innocuousness was held up to critical scrutiny early on in the interviewing.

### Interviewers as seekers and purveyors of knowledge

Seemingly mundane research questions can disrupt household power relations. Take, for example, this excerpt from a young mother's second interview describing her reaction to what I thought were a series of mundane, matter-of-fact, survey questions about household responsibilities:

> You know I remember filling out that checklist on your study's questionnaire of all the things that are your responsibility, and the only thing that I can remember not being my responsibility was the cars. That was overwhelming to me. And I remember checking it off and going 'yeah, yeah', and now when you look at [my husband's] survey maybe he had most of his checked off too. And that was our lack of communication, but I remember just feeling pitiful.[3]

Interviews in the *Family Fantasies* project were preceded by a questionnaire survey that focused on checklist information that I felt would take up too much time in face-to-face meetings. A methodological dilemma is raised by this kind of approach because, at one level, it 'clues in' participants on your research interests, and enables household members to discuss your topic prior to the interview. Couples can corroborate and decide upon how they want to present their shared lives. As an interviewer you may gain from this collective presentation, but you are also a purveyor of knowledge and sometimes, as the above quote suggests, specific questions produce new kinds of knowledge that changes relations between household members.

*Family Fantasies* is a critical feminist study of interpersonal relations, families, communities and urban space. My initial intent was not necessarily consciousness raising amongst participants, although feminist politics framed many of my questions. Certainly, with the project came my opposition to racism, sexism and patriarchy and a commitment to understanding how we may better the lives of young parents, but the last thing I wanted was to make people feel, in the mother's words, 'pitiful'. This example precipitated the first of several moments throughout the project when I stopped to consider the extent to which seemingly innocuous questions about daily routines were a form of intervention.

### Interview styles and formats

The ethical dilemma of consciousness raising and possibly driving a wedge between partners is further exacerbated by the methodological decision of whether to interview household members together or separately. In *Family Fantasies* we tried to do both, with the joint interviews comprising unstructured discussions about raising children and separate interviews focusing on an individual's daily round. To the extent that household management, decision making and recreation are practised and experienced jointly, then it is potentially useful to interview couples together. Factual questions about time management and responsibilities around the house, or questions relating to collective experiences, are usefully probed in discussion settings that enable the response of one family member to stimulate responses in others. Inconsistencies in individual responses or disagreements may enable consensus building or couples may agree to disagree. Valentine (1999: 71) points out that separate interviews may generate anxiety if a partner feels that he or she is being talked about and cannot manage the outcome or defend him or herself against unjust accusations. One way to overcome this problem is to tailor separate interview formats to the particular experiences of individuals. In *Family Fantasies* we asked participants to go over their time budget for the previous day during one-on-one interviews. After our group discussions (the two interviewers and the couple), participants were usually comfortable splitting off with an interviewer to elaborate upon aspects of their daily round, and, for the most part, this focus did not alarm other household members. Sometimes there was more informal group discussion after the one-on-one interviews.

Even with personal interviews structured around a discussion of day-before experiences, some participants voiced concern about privacy and anonymity. Statements beginning with 'I wouldn't want Bob to know but ...', 'If Maria got wind of ...' or 'This is between you and me, but I think ...' were common. We assured our participants that their responses were confidential and would be shared with no one, including their partners. Interview excerpts would be published anonymously, with the identity of persons and places hidden by pseudonyms.

The level of seriousness with which privacy was taken suggested some interesting interpersonal dynamics on what couples preferred to keep secret from each other. For example, Amy pointed out that her partner's help was important to her sanity and that he was very helpful around the house.[4] Jokingly, she asked the interviewer not to mention this to him, because she did not want him to stop. The following interview excerpt from Amy's partner, Chad, highlights how humour sometimes structures their relationship:

> Mm ... there's a million unexpected things that may come up, like, if the baby's sick ...
> Well tonight I would like to have seen the [Post Office] open a little longer cause I didn't
> get to deliver something. [Interruption from partner in the other room: 'You didn't get the
> box off?' to which he responds: 'I didn't! I was so tied up today dear'.] Yeah [laughs], she
> makes me do everything. Wait 'til all these anthropologists in the future see how much
> stuff you make me do!' [This last sentence is called out so that the partner can hear.]

Bantering of this kind may provide insights into couple dynamics. The communicative nuances of body language, interruptions and the use of humour or rhetorical quips can help uncover subtle, but clear, indications of how couples compromise and contest aspects of day-to-day life.

### Learning from private and shared interviews

The above example pointedly reveals a problem with interview contexts that are less than private. When research questions are geared towards highlighting pre-existing power relations and uncovering potential social inequities, separate interviews with prescribed privacy and anonymity are a preferred format for fuller disclosure of information. If interviewees doubt the confidentiality of a discussion, then it is unlikely that they will broach sensitive topics. Candid responses to questions about power relations must be guarded by assurances of anonymity. If privacy is absolutely necessary owing to sensitive questions, it might be appropriate to schedule interviews when partners are not around. When interviewing in families – particularly in small homes – it is sometimes impossible to maintain separate discussions. Lack of privacy influences the kinds of information a participant is willing to discuss. The following excerpt from a student interviewer's field logs suggests how external influences may affect an interview:

> Sandra is seventeen, a high-school student who studies at home.[5] She is engaged to her
> boyfriend, the baby's father. Next interview we need to take Sandra out for lunch or
> something. They live in a very small place in Barrio Logan, there's no privacy. The family is
> quite close-knit ... it was hard to get Sandra's perspective without some 'input' or
> comments from other members of the family.

Lack of privacy during separate interviews can silence participants, but it may also engender coercion if partners are able to listen in on conversations. During one interview, a young mother who was playing with the baby while her partner was cooking in the kitchen got very upset when he kept eavesdropping and wanting to comment upon her statements. She replied pointedly that 'I think I am portraying you fairly' and then chastised him for making his own dinner and forgetting about the baby's needs. As the interviewer sat helplessly, the situation deteriorated into conflict, with the mother berating the father for not taking more household responsibilities. Conflicts such as this are uncommon in interviews, but they offer insight into how couples negotiate and share their lives. It is also possible to learn something when a partner is silenced by lack of privacy, as suggested by the following field-log excerpt:

> This interview was conducted in Spanish in a small one-bedroom bungalow. Both Maria and José were present for each other's interviews. This did not seem to impede José's answers, but did appear to impede Maria at certain moments. In particular, in the question about how much control she has over household decision-making, Maria would not say in front of José. She also consulted with him on certain points … During José's interview, Maria sat across from he and I and looked rather bored; she stared languidly into space while José offered his answers. On the other hand, José seemed to be listening intently during Maria's interview.

Partner politics often incite dominant perspectives, but they may also suggest important contexts of compromise and change. For example, an interview with a pregnant teenager and her mother suggested some interesting changes in the power dynamics of the household:

> Josephine is the mother of Liz who is having her baby in 5–6 weeks. The only problem with the interview was that Liz insisted on sitting next to us during Josephine's interview and kept telling her how she had answered. As she continued to interrupt her mother I sensed a wee bit of friction between the two.    (student field log)

Later interviews confirmed that, in Liz's case, having a baby raised her standing in the household to the extent that she was empowered to express her opinions forcefully. Fernandez-Kelly (1994: 109) notes that in lower-income households 'the birth of a child often manifests a human capacity for affirmation'. Liz viewed having a child as an avenue through which to reach adulthood and respect in the family and her attempts to dominate interviews highlighted her newfound empowerment. In other interview contexts, dominant views may suggest authority rather than affirmation and empowerment. Privacy may simply not be possible, but we can nonetheless learn much about power relations from interpersonal dynamics between interviewees. Ann Oberhauser (1997) points out that interviews in participants' homes are important strategically because they have the potential for disrupting power hierarchies between researchers and respondents. In the next section I move from couple dynamics to highlight interpersonal relations between interviewers and cohabiting interviewees.

## Being there

In-depth interviews are always collaborative processes between researcher and researched, but it is often unclear how power relations are exercised. Some texts on qualitative research tell us that we must give up control and expertise to our participants (Ely 1994; Denzin 1997), but this is rarely how power is constituted in the field. It is important to recognize the politics of difference that constitute our relationship with interviewees. We are always part of the academy and scholarly research, and as we encounter and engage with couples we accept that we can never fully comprehend their shared lives. We cannot let go of one realm fully to embrace the relationships we seek to understand. Cindi Katz (1994: 66) describes this fuzzy and unstable position as a 'space of betweenness'. It is a position that is always political, sometimes cathartic and empowering for both interviewer and interviewee, but just as often it is frustrating and enervating. Some frustration may arise from discovering that interviewees have their own agendas.

### Coupled expectations and reactions

Participants agree to be part of studies for a variety of reasons, and their expectations of the encounter may differ significantly from yours, as suggested by this field log excerpt: 'After the interview was over, David told me that he and his wife belonged to an evangelical church … He invited me to join his church which I thanked him for and told him I would keep it in mind if I ever wanted to.' For the most part, the couples we interviewed were very accommodating and willing to talk about their experiences as expectant or new parents, despite their busy schedules. Interviews in the home were particularly useful, because participants were free to point things out and demonstrate how the home space helped or hindered their caregiving (see Fig. 5.1).

**Fig. 5.1** *Sharing in the lives of families*

Participants usually do not hold the needs of a project nearly as highly as you do and there are always forgetful interviewees to frustrate careful planning:

> *Peter and his wife had forgotten about the interview so when we arrived on Saturday morning they were not really prepared to talk to us. Because of this, it seemed that Peter's answers were a bit short and possibly not as well thought out or expanded upon as most of the other respondents, and the interview was a bit more rushed than normal.* (student field log)

In addition, individual household members do not always expect the same thing from an interview. Often, one member persuades a recalcitrant partner to participate. Sometimes, unexpected household changes add a dramatic twist to carefully planned research agendas. A particularly distressing moment occurred for a student and me one Saturday morning as we stood at the door of an apartment intent upon interviewing a couple with infant twins. During the first set of interviews, Carol confided to an interviewer that Will's job 'with the Navy is a great source of my anxiety'. Will was interviewed separately on the balcony of their apartment and the interviewer noted in her field log that 'I think in the next interview, we should concentrate upon how Will feels about being away from his wife and new twins'. Prepared to follow that line of discussion eight months later, I rang the doorbell. A sinuous man in Wranglers jeans and a denim shirt opened the door. He held a cowboy hat in one hand and a duffle bag in the other. 'Who the fuck are you?' he challenged, before we had a chance to introduce ourselves. Carol's voice was heard from inside the apartment: 'They're here to do the interview.' Will opened the screen door and elbowed his way between us with an 'I'm out of here' shout over his shoulder. Carol came to the door carrying one of the twins. She was very apologetic: 'That was Will, I'm so sorry but he just left me, do you still want me in the study?' I said yes, but suggested that we could schedule another time. She insisted that it was no problem and so I proceeded with my questions while my colleague played with the twins. Will's name did not crop up during the interview nor did I feel inclined to raise the issue of his leaving. Although nothing was said, the incident at the door coloured the interview experience. It was clear from what she said that Carol was intent on a life as a single mother.[6]

Interviewee reactions sometimes took me by surprise. For example, on several occasions interviewees took offence at our line of questioning because of the ways in which they felt we reinforced essentialist notions of parenting:

> *Raoul really enjoys being a father, and is one of the few fathers who insists that if he had*
> *the opportunity, he would stay at home and take care of the baby. In fact, he did not like*
> *some of the questions because to him, they assumed that the mother is the only one who*
> *has the option of staying home and taking care of the child.   (student field log)*
>
> *Christie was a little upset about the interview because she felt that it was sexist in some*
> *way because it did not view home work as 'paid' employment. She viewed work as*
> *something that should be considered as equal as paid employment – and so for her the*
> *interview was a little offending.   (student field log)*

When these surprises occurred, we shared with interviewees the political assumptions and biases of the study and made changes where appropriate. And so, one year later:

> *Christie and her child were both very friendly. Christie seemed willing and eager to discuss*
> *anything at all; she was interested in the research. An earlier interviewer had noted in his*
> *field log not to imply that Christie did not work, for she considers raising her child a full-*
> *time job. We didn't have a problem this year.   (student fieldlog)*

The point I am trying to make here is that interview contexts are fluid and collaborative, sometimes requiring you to deviate from a carefully planned schedule. There is also nothing wrong with being proactive. Some participants expect you to elaborate on how their responses match up to those of their partners or others in the study. How do you respond to requests for information on 'how are we doing?'? I usually sidestepped specifics by offering data from the larger questionnaire survey that accompanied the study. If this did not satisfy, I would note that all caregivers use different strategies and that raising a child is profoundly complex. I would also share anything that I knew about available community and parenting resources if information was requested. Was this unduly influencing the interview and biasing my results? Absolutely!

The focus in this section is on 'being there' as it is embodied in relations between interviewers and cohabiting partners. Interview contexts often boil down to issues of trust, and, although aspects of anonymity can be assured prior to the interview, and privacy can be established during the interview, trust is often derived from how you position yourself politically in relation to participants.

## Personal politics and positionality

When interviewing partners who share aspects of their lives together, or when dealing with personal or sensitive issues such as illness, bereavement or the raising of a child, openness and honesty are basic requirements. Without honesty you rarely gain the trust of your participants. This does not mean that you should compromise your personal politics, but neither does it mean that you insensitively blurt them out in the middle of an interview. It is sometimes better to confide your politics to your field log:

> I interviewed a couple who lived in a middle class, predominantly white neighborhood of San Diego. When I asked the father about access to services and public transportation ... He stated that he feared that Mexicans would use the trolley to access his neighborhood and break into his home ... I felt his comment was unfair and racist. While I did not comment at the time, the attitude ... confirmed to me a less desirable characteristic of [white] Southern Californians more generally, which I also witnessed on other occasions.

It is impossible not to judge and so we need to think critically not only about how personal moral assessments colour interview experiences but also how we deal with those experiences. There are circumstances, thankfully rare, where your moral judgement and legal position may require you to compromise assurances of privacy and anonymity: 'While we were doing the interviews, the parents were playing with their son and the other interviewer and I felt uncomfortable with the roughness of the play. We couldn't wait to finish the interview.' (student field log). Over the course of multiple interviews with 120 participants this was the only encounter where we talked about calling California's Child Protective Services or some other authority because of suspected child or partner abuse. We decided in this case that the interviewers' discomfort did not indicate abuse, because the child did not seem in distress or at risk, but these are sometimes extremely difficult judgement calls. In the USA you are legally required to report abuse and I would not have hesitated if a child or partner was perceived to be at risk. The importance of action under those kinds of circumstances, although difficult, is mandated for me by concerns that go far beyond legalities or the academy. No one can predict how they will feel about encounters with abuse and most interviewers are not trained to deal with these kinds of circumstances. It is nonetheless useful to reflect honestly on how you might react if you suspect abuse. Moral assessments and ethics join project and personal politics in complicated ways, but one useful technique is to search underlying motivations because those are what drive positionality.

### Voyeurism and empathy

At first, I fixed the couples encountered in *Family Fantasies* with a gaze that was thinly veiled voyeurism. This voyeurism came in part from my background in cognitive and behavioural science that stressed value-free detached and distanced study. Although I was trying to unshackle myself from this background, the presumptions of its training wore on me heavily. Trust and the assertion of commonalties rather than the maintenance of an aloof academic position facilitate understanding. This occurred for me in the *Family Fantasies* project when I began sharing similar issues and problems that I had experienced as a parent. I am genuinely concerned about how we construct our identities – how they are linked to notions of family, community and society – and how we use these identities to position and understand people, places and events. My opposition to family systems rooted in patriarchal authority does not arise from a vague need to be politically correct but, rather, from a deep-seated anxiety about societal myths about fathering and my own sexual and social being. My children were 1 and 3 years respectively when the *Family Fantasies* project began and I wanted to understand how other parents cope with their political identities around children and partners. At that time, my partner and I often experienced difficulty communicating our needs and expectations around gender roles and relations as our daily rounds whirred chaotically between home and work responsibilities. In areas where my own family was not working, how could I learn from families similar to mine? With the realization of this underlying motive, my voyeurism turned to empathy and my interviews from question–answer sessions to discussions where I increasingly balanced the politics of my academic-self with the politics of my parent-self.

Implicit in what I am saying is that, as interviewers, we recognize that our identities are multiple and complex. We are more than researchers with clipboards and tape recorders, and the people we interview lead plural, complex, shared lives, much of which is hidden. Some of what we encounter may be disagreeable and in extreme cases sufficiently reprehensible to require some form of intervention on our part. What I am suggesting here is that, at the very least, our task is always to search within ourselves and to be very sure of our motivations so that we may act responsibly. Responsibility is about accommodating difference and this begins with recognizing otherness within ourselves and, by so doing, we begin to reach out with empathy to our interviewees.

## Morality plays: a concluding comment

When you embark upon a project that involves interviewing people who are cohabiting partners, then you become part of the life that they share, no matter how short and seemingly negligible your contact. The partnership may continue unaltered when you leave or it may continue in a different form, but your relationship with the couple raises important moral questions. Your impact on interviewees may be negligible or it may be foundational and there is power in recognizing the politics of difference that constitute your specific relationship with a couple or family. Current interests in the thorny and often intractable questions posed by feminism and poststructuralism demand methods in which we not only highlight our own subject position but also raise the importance of what Iris Young (1990) calls a 'politics of difference'. Simply put, Young's notion of plurality recognizes the power of space as an active participant in the way people love, ignore, cajole, tolerate, expect, consent, compromise, disagree, claim, come together or isolate. This power articulates morality in a form that recognizes individual responsibilities (to partners, children, communities and interviewees) and the overarching and sometimes overbearing influences of societal strictures.

I want to note in closing that there is as much power in setting aside your systematic search for knowledge (which is always biased by protocols and training) and trusting the interview process itself. This is not necessarily a breach of the scrupulous ethics laid down by Human Subjects Review Boards, but rather an admission of your humanness, your need to participate in conversations and your need sometimes to abandon the structure in your interviews so that discussions can evolve in unexpected ways. By so doing you become bricoleurs, ready to use anything that comes to hand, whether that be changing a diaper (nappy), informing someone about a local activist group, taking a grandmother out to coffee, or driving someone to the grocery store. The unexpected found in ourselves and in those with whom we connect can then be mapped, welded or sutured onto our research in ways project proposals can never anticipate.

## Notes

1. Over the course of four years, the *Family Fantasies* project followed 120 household members from prior to the birth of a first child to the child's third birthday; 95% of the interviews took place in participants' homes. The questions that the project set out to address focused on changing gender roles and relations around the birth of a first child. The interviews pivoted on household members' day-to-day activities (we started each interview with a time budget for the previous day), responsibilities, feelings, attitudes and opinions about family life. The interviews created qualitative information on the interdependencies and relations between household members in terms of changing commitments to work and domestic realms. Names of pregnant mothers were obtained from a local Health Maintenance Organization and the mother always acted as a gatekeeper. To our knowledge, no couples in the study were gay or lesbian. Pseudonyms are used in this chapter to protect participants, and the people in the introductory photograph are not related to any of the stories presented.

2. Over the life of the project, 14 students were employed to help with interviewing. For the most part, both myself and a student or students in pairs conducted interviews. After each interview we wrote about the encounter in unstructured field logs. The use of field logs or field diaries is a fairly common approach employed by ethnographers to document their experiences and feelings (see Ely 1994). Field logs not only provide a valuable context for interview transcriptions but sometimes, as in this case, they may be used as sources of primary data.

3. A fuller version of this story and the outcome of this young mother's feelings of 'pitifulness' are found in Aitken (2000).

4. I elaborate on Chad and Amy's conflict over the work of parenting in a paper about how some 'rites of passage' are embodied in the birth of a first child (Aitken 1999).

5. Sandra's story is elaborated – primarily through the voices of her father Benito and her Uncle Roberto – in Aitken (1998).

6. Carol's story is elaborated upon in Aitken (1998).

## Acknowledgements

Research for this chapter was supported in part by Grant SES-9113062 from the National Science Foundation and a grant from San Diego State University Foundation. Special thanks go to the students who were employed on the study: Marta Miranda, Chris Carter, William Granger, Leslie Bolick, Suzanne Michel, Thomas Herman, Katina Pappas, Nickolas Deluca, Meg Streiff, Susan Mains, Michael Cohen, Matt Carroll, Serena McCart and Pauline Longmire. In addition, I would like to thank all the families in San Diego who filled out questionnaires and/or agreed to be interviewed as part of the study. Any opinions, findings and conclusions or recommendations expressed in this chapter are mine and do not necessarily reflect the views of the National Science Foundation, San Diego State University or the students and families involved in this project.

## Key references

Ely, M. 1994: *Doing qualitative research: circles within circles.* London and New York: Falmer Press.

Kvale, S. 1996: *Interviews: an introduction to qualitative research interviewing.* Thousand Oaks, CA: Sage.

*Professional Geographer* 1994: *Women in the field: critical feminist methodologies and theoretical perspectives.* Special focus section. 46(1), 54–102.

Valentine, G. 1999: Doing household research: interviewing couples together and apart. *Area* 31(1), 67–74.

## References

Aitken, S. C. 1998: *Family fantasies and community space.* New Brunswick, NJ: Rutgers University Press.

——— 1999: Putting parents in their place: child rearing rites and gender politics. In Teather, E. K. (ed.), *Embodied geographies: spaces, bodies and rites of passage.* London: Routledge, 104–25.

——— 2000: Mothers, communities and the scale of difference. *Journal of Social and Cultural Geography* 1(1), 69–86.

Allan, G. 1980: A note on interviewing spouses together. *Journal of Marriage and the Family* 42, 205–10.

Denzin, N. 1997: *Interpretive ethnography: ethnographic practices for the 21$^{st}$ century.* Thousand Oaks, CA: Sage Publications.

Elwood, S. and Martin, D. 2000: 'Placing' interviews: location and scales of power in qualitative research. *Professional Geographer* 52(4), 649–57.

Ely, M. 1994: *Doing qualitative research: circles within circles.* London and New York: Falmer Press.

Fernandez-Kelly, P. 1994: Towanda's triumph: social and cultural capital in the transition to adulthood in the urban ghetto. *International Journal of Urban and Regional Research* 18, 88–111.

Goss, J. 1996: Introduction to focus groups, *Area* 28(2), 113–14.

Handel, G. 1996: Family worlds and qualitative family research: emergence and prospects of whole-family methodology. *Marriage and Family Review* 24(3), 335–48.

Hertz, R. 1995: Separate but simultaneous interviewing of husbands and wives: making sense of their stories. *Qualitative Inquiry* 1(4), 429–51.

Katz, C. 1994: Playing the field: questions of fieldwork in geography. *The Professional Geographer* 46(1), 67–72.

Longhurst, R. 1996: Refocusing groups: pregnant women's geographical experiences of Hamilton, New Zealand/Aotearoa. *Area* 28(2), 143–9.

Oberhauser, A. 1997: The home as 'field': households and homework in rural Appalachia. In Jones, J. P., Nast, H. and Roberts, S. (eds.), *Thresholds in feminist geography*. Lanham, MD: Rowman & Littlefield, 165–82.

Sparke, M. 1996: Displacing the field in fieldwork: masculinity, metaphor and space. In Duncan, N. (ed.), *Body space: destabilizing geographies of gender and sexuality*. London: Routledge, 212–33.

Valentine, G. 1997: Tell me about . . .: using interviews as a research methodology. In Flowerdew, R. and Martin, D. (eds.), *Methods in human geography: a guide for students doing a research project*. London: Longman, 110–26.

—— 1999: Doing household research: interviewing couples together and apart. *Area* 31(1), 67–74.

Young, I. M. 1990: *Justice and the politics of difference*. Princeton, NJ: Princeton University Press.

# 6

# Cross-cultural research: issues of power, positionality and 'race'

## Tracey Skelton

## Introduction

While I was at junior school and 8 years old, I remember vividly a visit from Miss Peacock. She was a missionary, a white woman with grey hair fastened in a bun. She told us about India and her work there, and she showed us how to put on a bright blue sari. This woman fascinated me. Here she was, in ordinary clothes looking so typically 'English', and yet when she wore the sari and used Indian words she was someone else, seemingly part British and part Indian. What amazed me most though was that someone could go and live in another country, spend time working there and learn about another way of life (Miss Peacock had been in India for more than twenty years). Miss Peacock was the first person I had ever met and talked to who had lived abroad. I remember walking home from school dreaming of the chance to spend time 'abroad', to live somewhere completely different and to learn another way of life.

Fourteen years later I embarked on my Ph.D. study. I chose the advertised project because there was a key focus on gender relations and also there were three options offered for 'doing the fieldwork': the north-east of England, Mexico or the Caribbean. My childhood dream of going overseas had been reinforced through studying anthropology as part of my first degree; the desire to study and conduct research from a feminist perspective emerged through my degree and by getting involved in feminist activism. Doing a research project on gender relations with an overseas fieldwork component was the ideal combination of my two wishes. I chose to conduct research into gender relations in the Caribbean, specifically on the island of Montserrat.

My research had an explicit focus on feminist theory and explored gender relations in four areas of social organization: the household, the workplace, sexual union patterns and heterosexual behaviour drawing on the concept of patriarchy (see Foord and Gregson 1986). The research aimed to explore the complexities of power between men and women on Montserrat through qualitative methods. Hence I used semi-structured, interactive interviews with women from all over the island, and very long, unstructured interviews with men and women within the village where I lived (see Fig. 6.1). I wanted to investigate the ways in which patriarchal gender relations were created and reproduced – how was gender power maintained and to what extent was it

**Fig. 6.1** *Tracey Skelton with two of her Monserratian neighbour/interviewee's children*

challenged and circumvented? In the semi-structured interviews I asked questions that focused on the households in which interviewees had grown up and the ones they currently resided in. This demonstrated the fluidity and dynamism of household structures, the prevalence of extended and three-generational households and also the significant role played by many women as heads of household. The next part of the interview focused upon the workplace: people's work histories, wages, whether this met their needs and how they developed coping strategies if necessary, and what their employment aspirations were. In this way I was able to gather information on gender roles and expectations within paid employment and also to analyse the ways in which gender was played out both in the workplaces and also through the experience of having waged work. I talked to women about children, their experiences of having children, whether they wanted to have children and why, and this led into a discussion of contraception and abortion (an extremely difficult subject to explore). Having discussed these more intimate parts of women's lives and built up some degree of rapport with the interviewees, I then moved into the final section, which focused on sexual relationships including questions about domestic violence, marriage, cohabiting, 'visiting' relationships, concepts of power and how power was maintained. The research was feminist both in its theoretical focus – the nature of the questions and the themes it explored – and in the possibilities for sharing information with women, in particular the legal rights they had, which I learned from my research on the island's legislation with legal officials and the police.

My fieldwork time was initially about my wishes, dreams and desires. However, during my time in Montserrat (and subsequently), the sense of responsibility, the questions of representation (Radcliffe 1994) and the intense dilemmas relating to cross-cultural research (Mullings 1999) began to emerge. These complexities and dilemmas later came to haunt me and paralyse me, and only recently have I gained the confidence to begin to engage with complex questions and personal anxieties about my research work in the Caribbean within a semi-public domain.[1]

## The dilemmas and complexities of cross-cultural research

In this chapter I want to draw on my personal experience to discuss some of the issues that arise when doing what we might call 'cross-cultural' research. Cross-cultural research can take place anywhere. The 'cross-cultural' element comes, not from moving places, but when research takes place between people of different cultural heritages, backgrounds and practices. In many cases, as through my work in the Caribbean, and much more recently with UK-based Montserratians who have been dislocated owing to a volcanic eruption, cross-cultural research is also 'cross-racial'.[2] Almost everyone I have conducted research with in the Caribbean is black, of African descent, and I am white, of European descent.[3] What I cannot do in this chapter is explore all the complexities and issues that arise in relation to conducting cross-cultural research, especially when complexities of 'race' are involved. However, there are particular configurations of power that frequently emerge in cross-cultural and cross-racial research. This chapter considers some of the most significant, but is by no means a definitive discussion.

It is crucial in any research that we consider our positionality and what that might mean in relation to the ways in which we do our research, and how the people we work with perceive us. By positionality I mean things like our 'race' and gender as discussed above, but also our class experiences, our levels of education, our sexuality, our age, our ableness, whether we are a parent or not. All these have a bearing upon who we are, how our identities are formed and how we do our research. We are not neutral, scientific observers, untouched by the emotional and political contexts of places where we do our research. We are amalgams of our experiences and these will play different roles at different times. If we work in a post-colonial geographical context, then being white and born in the former colonial country may have an important impact upon the relationships we can establish during our research. It might mean that people feel they have to talk to us out of an encultured sense of deference; it might mean that people do not want to talk to us, because we represent a very negative and exploitative past. However, it might also be that people who live in the areas where we are planning to do our research decide to keep an open mind about us as individuals, wait until they get a sense of who we are and what we are like and then choose whether or not to get involved with the research. Nevertheless, part of our honesty and integrity as researchers must be based upon considerations about ourselves, our positionalities and our identities and what role they might play in our research. We should also recognize that the research experience may (and I would argue should) bring about changes within ourselves too.

Doing cross-cultural research is full of complexities and contradictions, which may at times mean we have doubts about whether we should do the research at all. However, I would argue that such doubts and dilemmas are good and a productive part of effective and sensitive cross-cultural research. It is also useful to know that, as we go through these anxieties, we are not alone. These research and writing dilemmas are not only present within cross-cultural research but are also embedded in feminist research. Diane Wolf opens her introductory essay to *Feminist Dilemmas in Fieldwork* (1996) thus:

> Feminist dilemmas in fieldwork are [ethical, personal, academic and political]; they gnaw at our core, challenging our integrity, our work, and at times, the raison d'être of our projects. Feminist dilemmas in fieldwork revolve around power, often displaying contradictory, difficult, and irreconcilable positions for the researcher. Indeed, the power dimension is threaded throughout the fieldwork and post-fieldwork process and has created a major identity crisis for many feminist researchers. (Wolf 1996: 1)

We could insert after 'feminist' the words 'cross-cultural', because, when such research involves questions of 'race', researchers can go through very similar crises.

From what I have written above the reader may be forgiven for wondering if the complexities and difficulties of cross-cultural research are worth it! Personally I think they are, but there are a great many things to think about before, during and after we conduct such research. The existing literatures on feminist dilemmas in fieldwork and research are extremely valuable in helping us work through the problematics of doing cross-cultural research well. By 'well' I mean with a politics that: acknowledges, respects and works with difference; recognizes and takes responsibility for differential power relations that may exist between the researcher and those participating in the research; chooses methods that empower the 'researched' and that allow depth of analysis and complexities to come forth; and challenges and transforms unequal power relations (for more detailed discussion of politics in feminist geography research methodologies, see McDowell 1992; Nast 1994; Staeheli and Lawson 1994; Madge *et al.* 1997; Raghuram *et al.* 1998).

Let us think through some examples of these. To acknowledge, respect and work with difference mean first that you have to: recognize the many differences that exist between yourself as researcher and those you want to work with; reflect upon what they might mean; and then think through how you deal with those differences and make them part of the research, rather than ignoring them. This is also connected to the important recognition of power in the research process. At the time of my Ph.D. research I was in my early 20s. At this age most women in Montserrat have had at least one if not two or three children. I was not a mother and this was therefore a marked difference between myself and almost all the women I interviewed. It was a common question asked of me by the women and men, and they also wanted to know why not. Rather than trying to cover up this difference or feeling upset by intimate questions, I was honest about my reasons for not having children and talked through the ways in which it made us different. In Montserratian culture, motherhood is extremely important and is the most clearly recognized way of girls becoming women. Hence it is a mark of status, and I did not have it. In the context of the interviews, this meant that some of the women felt luckier, more mature or more powerful than me. I found this a healthy way of letting the power I had in the interview context – I was the one asking the questions – dissipate and shift into complex positions within the interviews. It was part of the 'negotiations'[4] that took place throughout each interview.

There are several possible ways of using methods that empower those we work with. Allowing people to 'opt into' research is very important and this can be a genuine 'opting in' only if people know a considerable amount about the research project. Hence I always take time in interviews to explain any research project I am working on, leave time for questions and give people the chance to change their minds after the research has been explained. It is also important to introduce our projects in ways that people will understand and can relate to.

The vast majority of people who agree to give interviews do it in an altruistic way; they get relatively little out of it, but they give us an awful lot. For my research I favour interviews that allow the space and time for people to talk about things they want to tell me. Open-ended questions let them emphasize and omit particular things, to reflect on their past, to impress you with their knowledge. As an interviewer, I consider it is important to let people know, repeatedly, how useful what they say has been. Some of the people I worked with in Montserrat would tell me that they had little to say because they lacked 'education'. After some of the richest and deepest interviews I have ever been privileged to have been part of I have assured people that what they lack in formal education is more than compensated for by their depth of knowledge and their skills in articulating that knowledge. I have then heard these same people telling work mates or family members that they were 'articulate', 'fluent' and had 'lots of knowledge'. Such an interview experience is empowering for both people. Allowing the space for people not to answer questions and respecting their decisions as soon as they refuse to talk about something are also about respect and empowerment; they give the interviewee some control within the interview.

What I want to do in the remainder of this chapter is examine some of the issues that have to be considered when doing cross-cultural fieldwork.

## Questions of positionality and power

There are extensive critiques of academic feminist theory that emerged in the UK and the USA through the 1970s and early 1980s, which take feminism to task about its ethnocentrism, its almost exclusive focus on white women's experiences and its denial to consider its own role in racist processes, including the ways in which feminists conducted research with 'black' and Third World women[5] (Carby 1983; hooks 1982, 1984, 1991, 1992; Collins 1990; Mohanty 1991; Patai 1991). Building upon these critiques and through constructive dialogue between feminists of different backgrounds, geographies and histories, there is a wealth of literature that has provided the conceptual tools to interrogate both gender and 'race' (Afshar and Maynard 1994; Kobayashi 1994; Kobyashi and Peake 1994; Marshall 1994; Phoenix 1994; Skelton 1995, 1999; Anderson 1996; Wolf 1996; Dwyer 1999). Indeed it is such developments within feminist academic debate that have freed many researchers from a 'paralysis of race'. This paralysis had prevented genuine articulation of research considering the intersections of 'race' and gender. Subsequently, in feminist geography particularly, and human geography in general, there have also been sophisticated debates about the concepts of 'insider'/'outsider' roles and power in our research (Gilbert 1994; Kobayashi 1994; Herod 1999; Mullings 1999).

A key facet of feminist debate about methodologies has been the importance of recognizing the roles positionality and power play in our research, and this is essential in cross-cultural research. As a young white woman embarking on Ph.D. research, I had never been encouraged to think about my racial identity; I knew I was white, but I had not been forced or encouraged to think about this in relation to my research project. Politically I was more in tune with 'race' through an active involvement in anti-racist activities while in my sixth form and through my undergraduate years. Through my political action I thought that I had worked through issues of 'race', but in reality I knew more about racism than I knew about my own racial identity and what that could mean in the context of the Caribbean. Such naivety makes me cringe now, but did I learn fast!

In Montserrat, particularly in the village where I lived for just under a year, my whiteness was constantly referred to, admired, joked about, criticized and stared at. I was reminded that I was white daily and so forced to recognize that my whiteness had significant meaning in Montserrat. Once on Montserrat I learned that white people in Montserrat were one of four groupings. These included rich ex-patriots from the USA and Canada who had bought land in designated areas of the island and built luxurious homes; second, employees of the British government; third, staff or students at the American University of the Caribbean pre-medical school; and, fourth, development volunteers (US Peace Corps or British VSO). As a lone, British, female researcher I did not fit into any category, and yet for a long time Montserratians I met in the capital town of Plymouth and elsewhere would assume one of these roles for me. During my time in Monteserrat I had time for my role to begin to make sense and to have the chance to get to know people, so that my research and 'independent' status became recognizable and accepted.

What I learned while in Montserrat, and also analysed more when back in the UK, was that my whiteness was always going to be a significant factor in my research in the Caribbean, that it was frequently going to be uncomfortable and that it might also cause problems. I learned to be constantly aware of the potential power being white in Montserrat gave me, but also to recognize the barriers it would form. Some people would feel duty bound to talk to me because I was white and British, others would refuse to speak with me for those two same reasons; some men wanted to talk to me simply because I was white and female, some women would not talk to me for exactly those reasons. Then, as personal relationships were established, choices were made to get involved in my research or not based on other factors – such as not wanting to be part of a research project, wanting to help me with my research, a desire to help a friend. Each interview was a negotiation, and complex facets of positionality and power (both mine and those of the person who might or might not be interviewed) came to play at different points.

In the context of interviews conducted in the village where I lived, I realized, afterwards, that I had been 'tested'. Initially two women agreed to give me interviews, during the first two weeks of my moving into the village. Then I could not get anyone else to agree to an interview. People did not refuse exactly, but rather put me off until later, cancelled arranged appointments, postponed interviews. Two months later people who had postponed interviews began agreeing to new times; indeed, they asked me to call round, taking the initiative themselves; other people I asked agreed. I talked this through with one of my friends in the village and she explained to me what had been happening. She told me that people had been waiting to see if I could be trusted, if the information given in interviews would be kept confidential. They waited to see if any of the stories and information given to me in the first two interviews 'leaked out' into the village. When it was clear that no gossip was emerging from me, then it was proven that I could be trusted with interviews.

In the workplace-based interviews, in which I interviewed women employees who came from all over the island to their places of work in the Plymouth (capital-town) area, there were a range of ways in which I came to interview women, and this depended upon the management approach of their employers, all of whom I had interviewed first to get permission to interview employees and to gather contextual information about work. In some cases the manager agreed to a certain number of interviews and would select people, who would be sent into me. I found this an uncomfortable process of selection, because I had no idea whether the women had been given the chance to refuse to participate. In the first factory and on the first day I did interviews the women had not been given an option. At the outset of the interview I emphasized that they did not have to participate and gave a very clear and detailed description of what I was doing. After this all of the women said they were happy to continue.

These examples emphasize that I gathered my research only because Montserratians agreed to give it to me. I was a disempowered researcher in that context, because without consent to interviews I could not do the research. Nevertheless I knew that I had power invested in me because I was white, a full British citizen, educated and an outsider. I did not forget this power, but worked hard at countering it, because I did not want stereotypical assumptions to be made about who I was and what I thought, just as I was at pains not to do this about other people. So I would be vocally critical of the British government and talk through politics relating to colonial histories and presents and I articulated my thoughts about racism in British society. For me these conversations were as much a part of my research as formal interviews. As people learned more about my politics, and what I valued in life, then more people agreed to give interviews. However, what also became apparent through these 'talking times' was that my time in Montserrat was much more than time to do research: it was about a cross-cultural relationship developing that would change my life and my politics.

## Planning cross-cultural research

Having talked through some of the more esoteric and deeply personal aspects of cross-cultural research that we cannot plan for but that certainly open our eyes, let us consider aspects of preparing for cross-cultural research. It is extremely difficult to offer any definitive advice, because each research context will be different and the combinations of 'cross-cultural, cross-racial' research are enormous. Being black and doing research with white people will raise very different questions from those I have thought through in relation to being white and doing research with black people, for example. The same is true for research and researchers across the 'Third World/First World' boundary. However, I think there are some general principles that we all have to think about in relation to such research.

An important starting point is to find out as much as possible about the place and/or cultural context into which you are planning to take your research. However, be critical of what you read. Much of what I read about the Caribbean written by male, white 'First World' academics had distinct problems about lack of 'race' awareness and little if any discussion about positionality. I also found that much of it was condemnatory in tone, critical of the so-called dysfunctional family formations in the region (Henriques 1949; M. G. Smith 1963; Goldberg 1976; R. T. Smith 1988). Try and read as much as you can written by people from the culture/region yourself – I found Caribbean novels extremely valuable, especially those written by women. Think too about the ways in which you, your positionality, your historical and geographical contexts connect with the place/culture you are going to work in. Colonial relations remain significant in contemporary times; think through what such histories and politics might mean for your own research. Just because as an individual you were not part of colonial practices does not mean they are not part of our identities, the ways in which we see the world, or the ways in which the world sees us.

Think carefully about why you are doing the research you are planning and where you are doing it. Be honest about your reasons; accept that these might change over time and may well be fundamentally challenged by those who you do your research with. Face the dilemmas of conscience that will, and indeed should, emerge through this type of research; try and work your way through them, and read material written by those who have faced them. There is no real end point to the discussions and complexities surrounding cross-cultural fieldwork. At certain stages you will learn something new; there will be 'places' where you will be unhappy and find it impossible or extremely difficult to do any work; but there will be points where you feel that you have 'arrived', that everything falls into place – these will be (and should be) transient stages, but nevertheless can enable a very productive time.

Once you have thought through and come to some tentative conclusions about your positionality, think carefully about the research methods you will use and why. Think about whether you have chosen them because they are the best for you and your project, or whether you have thought about what might be best for the people you are working with. For my own cross-cultural research I have used semi-structured, interactive, intensive interviews, because I value the depth and richness that ethnographic research methods can provide. Such depth, experience and personal stories can really be collected only through qualitative methods. However, such methods also have their own critiques. The key thing is to identify which methods are going to be most appropriate to gather the information you are seeking. Some people might be happier filling in an anonymous questionnaire about very private things (Madge et al. 1997).

## The crisis of production

In this final section I discuss the dilemmas I have faced since conducting cross-cultural research. This relates to the very complex processes of reflection and writing up. There are several problems to be faced in trying to find the balance of not appropriating or speaking for those we have worked with but at the same time in some way telling their stories in their own words.

There are often crises of production in the 'post-fieldwork' context. I have faced such crises myself and made a difficult decision in the early 1990s to postpone publications from my Ph.D. thesis until I was sure I had resolved the dilemmas. My stance now is that, as part of the politics of reflective and politically conscious feminist and/or cross-cultural research, we have to continue our research projects, we must publish and disseminate our research. If we do not, others without political anxieties and sensitivities about their fieldwork processes take the space and may perpetuate negative representations and stereotypes.

In our writing we need to examine the complexities of how we write, what and who we represent and how. We must be constantly vigilant about stereotypes and assumptions that we as the writer may perpetuate and that our readers may interpret. This does not mean being dishonest about our research findings, but we have to think carefully about ways of writing that do not feed straight into racist structures. For example, in Monserrat I asked questions about domestic violence. It was clear that there were cases of this, some of them extremely violent. Initially in writing my thesis I was concerned that, if I recorded such information, I would be participating in a pervasive stereotype of the 'violence of black men'. However, women had told me their stories and the issue of violence is highly pertinent to gender relations and constructions of power, central themes within my research. I put this information into my thesis (Skelton 1989) and subsequently presented a conference paper on the subject, but it was framed in a detailed social context. I made it clear that the rates of domestic violence in Montserrat were not higher than the rates per capita in other parts of the world, talked through police and institutional responses to domestic violence and also wrote about the ways many women had articulated resistance to the violence. I also included male voices from the research who had talked against men using violence against women. In this way nuances and dialogues that occurred within the Montserrat context were represented to show that the situation is as complex and dynamic as it is, for example, in the UK. This, therefore, is part of writing with responsibility and attention to the context in which our research will be received.

## Conclusion

Cross-cultural research is difficult, particularly if we think through and acknowledge the complexities, sensitivities and dilemmas that are implicated within it, but that does not mean that we should abandon doing it. What it does mean though is that we have constantly to think about what we are doing, why we are doing it and what the research we do means to other people. I began my research in Montserrat because of a childhood dream and feminist politics. I continue to do research in Montserrat because of the relationships I have built up with Montserratians and because I can use the material and academic resources to which I have access to make sure that sensitive research is conducted with Montserratians. If I am truly honest, it is also because I love the place and many of the people there. I also enjoy who I can be when I am in Montserrat. I cannot abandon the history and relationships I have with the place and the people because doing cross-cultural research responsibly is hard.

Finally, as you think about the research you are planning, think too about what you will gain, and have gained, from your research. I do not mean another publication or qualification; I mean as an individual with a unique opportunity to spend time in another place/culture with a specific purpose to learn from other people. It is an extremely rare and privileged opportunity. My times on Montserrat have been some of the best of my life. When we are offered such unique and positive opportunities, it is important that we use them responsibly, with a politics that is about empowerment and equity and from which we produce material that challenges unequal power relations as much as possible. What our work should not do is reinforce and perpetuate racist stereotrypes; it should rather be part of challenging such stereotypes and damaging representations. In different cultures we might need to produce our work in different ways. In Montserrat people listen avidly to the national radio station, Radio Montserrat. Each time I have been back to the island I have been asked if I might do a live interview about the research I am planning to do, or talking about the outcome of a previous project. This lets people know that you were successful in your own project and is a way of publicly thanking people.

For me, my positionality and the cross-cultural research I do are very much part of a politics about the recognition of power and trying to challenge existing power relations. It is not always easy, as we have to live with many contradictions, but I think it makes for interesting research, and certainly research that is rewarding for others and not just ourselves. Our work might not be able to change the world but it is important that it does no further damage or violence.

## Notes

1. I say semi-public domain because I mean the presentation of seminars within university establishments and now this written chapter. I have given four seminars relating to my research work, positionality, ethics and 'race'. Three of them were to the gender and development units and women's studies groups based on the three campuses of the University of the West Indies in Jamaica, Trinidad and Barbados. The feedback and discussion from Caribbean women academics was invaluable. I only wish my own fumbling writings could do justice to the clarity of their words.

2. I have tried in vain to find a term that will work better than 'cross-racial' and yet still explain what I want to talk through in this chapter. I appreciate that there are problems with the term; for one thing it reminds me of the biological context of cross-pollination. It also remains locked into the conceptual idea that there are 'racial' differences, which stems from social Darwinism and scientific constructions of racism. However, although there is no genetic evidence to prove that there are different races (the genetic variety within a same skin-colour group is greater than that which can be found between two so-called different racial groups), nevertheless 'race' remains a significant social and cultural construct. I use the term cross-racial throughout this part of the chapter to mean a case where the 'race' of a person doing research is different from that of the people, or the majority of the people, who are participating in the research. Later in the chapter I refer only to cross-cultural research, because the repetition of the two terms cross-cultural and cross-racial can become irritating for the reader. I am henceforth using cross-cultural to mean research that takes place both across cultures and across 'race'.

3. I have conducted four significant periods of research in Montserrat and have travelled to the island two other times for social visits over a 15-year period. In the combined times of interviews I have interviewed only four white people; three were British government officials appointed to work for the Foreign and Commonwealth Office or the Department for International Development, and one was a woman who began work on Montserrat as a development volunteer, married a Montserratian and remained on the island for more than 30 years.

4. By negotiations I mean the kind of weighing up that always goes on in an interview. As an interviewer, in your mind, you might be thinking will this person tell me personal details? Can I actually ask the question about abortion? Does this person really want to be here or did her employer make her? Will she be offended if I ask her about sexual relationships? and so forth. All the time we are actively thinking about the discussion, the next question, whether we are clear as we speak and so forth. The interviewee will have other thoughts in her mind: is this the kind of thing she wants to hear? I wonder if there is a right or wrong answer to this? Will

she think this answer is foolish? How can I tell her I don't want to answer that question? What will she ask me next? Can I trust her with this information? Hence we effectively both negotiate our ways through interviews and many of the interviews are compromises. As interviewer you want to get the kind of information that will be useful, but you don't want to offend or upset someone, especially if you are a guest in her home, her community, her country. As an interviewee there is a sense of wanting to know you have helped but that you haven't been so open that you have put yourself at risk of embarrassment.

5. While I acknowledge there are complexities around the terms 'black' and 'Third World' women, I have chosen to use them as shorthand for other terms that would include women of the African and Asian diasporas, women of colour (as used in the USA), women of once-colonized countries and territories. Women of the Caribbean tend to identify politically as Third World women and on Montserrat the women would also self-identify as black. I do not use these terms to mask the diversity within such groups of women, nor do I intend to define them as a homogeneous groups. However, I have taken the terms of self-definition for many Caribbean women and use them to imply all women previously ignored or falsely represented by white Western feminism.

## Acknowledgements

My thanks go once again to the people of Montserrat, who have always helped me with my research and have taught me so much about experiencing and being 'cross-cultural' in all its manifestations. I would also like to thank Claire and Melanie, the editors of this volume, for their patience, helpful comments and advice.

## Key references

Kobayashi, A. 1994: Coloring the field: gender, 'race', and the politics of fieldwork. *Professional Geographer* 46(1), 73–80.

Madge, C., Raghuram, P., Skelton, T., Willis, K. and Williams, J. 1997: Methods and methodologies in feminist geographies: politics, practice and power. In WGSG (Women and Geography Study Group), *Feminist geographies: explorations in diversity and difference.* London: Longman, 86–111.

Mullings, B. 1999: Insider or outsider, both or neither: some dilemmas of interviewing in a cross-cultural setting. *Geoforum* 30, 337–50.

Patai, D. 1991: US academics and Third World women: is ethical research possible? In Berger, S., Gluck, S. B. and Patia, D. (eds), *Women's words: the practice of feminist oral history.* New York: Routledge, Chapman & Hall, 137–53.

Raghuram, P., Madge, C. and Skelton, T. 1998: Feminist research methodologies and student projects. *Journal of Geography in Higher Education* 22(1), 35–48.

Wolf, D. 1996 Situating feminist dilemmas in fieldwork. In Wolf, D. (ed.), *Feminist dilemmas in fieldwork.* Boulder, CO: Westview Press, 1–55.

## References

Afshar, H. and Maynard, M. (eds) 1994: *The Dynamics of 'race' and gender: some feminist interventions.* London: Taylor & Francis.

Anderson, K. 1996: Engendering race research: unsettling the self–Other dichotomy. In Duncan, N. (ed.), *Body space*. London: Routledge, 197–211.

Carby, H. 1983: White women listen! Black feminism and the boundaries of sisterhood. In Carby, H. (ed.), *The empire strikes back*. London: Hutchinson, 212–35.

Collins, P. H. 1990: *Black feminist thought: knowledge, consciousness and the politics of empowerment*. London: Hyman & Unwin.

Dwyer, C. 1999: Veiled meanings: young British Muslim women and the negotiation of differences. *Gender, Place and Culture* 6(1), 5–26.

Foord, J. and Gregson, N. 1986: Patriarchy: towards a reconceptualisation. *Antipode* 18, 186–211.

Gilbert, M. 1994: The politics of location: doing feminist research at 'home'. *Professional Geographer* 46(1), 90–6.

Goldberg, R. S. 1976: The concept of household in the East End, Grand Caymen. *Ethnos* 1, 116–32.

Henriques, F. 1949: West Indian family organization. *American Journal of Sociology* 55(1), 23–41.

Herod, A. 1999: Reflections on interviewing foreign elites: praxis, positionality, validity, and the cult of the insider. *Geoforum* 30, 313–27.

hooks, b. 1982: *Ain't I a woman: black women and feminism*. Boston: South End Press.

—— 1984: *Feminist theory: from margin to center*. Boston: South End Press.

—— 1991: *Yearning: race, gender and cultural politics*. London: Turnaround.

—— 1992: *Black looks: race and representation*. Toronto: Between the Lines.

Kobayashi, A. 1994: Coloring the field: gender, 'race' and the politics of fieldwork. *Professional Geographer* 46(1), 73–80.

—— and Peake, L. 1994: Unnatural discourse: 'race' and gender in geography. *Gender, Place and Culture* 1(2), 225–43.

McDowell, L. 1992: Doing gender: feminism, feminists and research methods in human geography. *Transactions of the Institute of British Geographers* NS 17(4), 399–416.

Madge, C., Raghuram, P., Skelton, T., Willis, K. and Williams, J. 1997: Methods and methodologies in feminist geographies: politics, practice and power. In WGSG (Women and Geography Study Group), *Feminist geographies: explorations in diversity and difference*. London: Longman, 86–111.

Marshall, A. 1994: Sensuous sapphires: a study of the social construction of black female sexuality. In Maynard, M. and Purvis, J. (eds), *Researching women's lives from a feminist perspective*. London: Taylor & Francis, 106–24.

Mohanty, C. T. 1991: Cartographies of struggle: third world women and the politics of feminism. In Mohanty, C. T., Russo, A. and Torres, L. (eds), *Third World women and the politics of feminism*. Bloomington: Indiana University Press, 1–47.

**Tracey Skelton**

Mullings, B. 1999: Insider or outsider, both or neither: some dilemmas of interviewing in a cross-cultural setting. *Geoforum* 30, 337–50.

Nast, H. 1994: Opening remarks on 'women in the field'. *Professional Geographer* 46(1), 54–66.

Patai, D. 1991: US academics and Third World women: is ethical research possible? In Berger, S., Gluck, S. B. and Patia, D. (eds), *Women's words: the practice of feminist oral history.* New York: Routledge, Chapman & Hall, 137–53.

Phoenix, A. 1994: Practising feminist research: the intersection of gender and 'race' in the research process. In Maynard, M. and Purvis, J. (eds), *Researching women's lives from a feminist perspective.* London: Taylor & Francis, 49–71.

Radcliffe, S. 1994: (Representing) post-colonial women: authority, difference and feminisms. *Area* 26(1), 25–32.

Raghuram, P., Madge, C. and Skelton, T. 1998: Feminist research methodologies and student projects. *Journal of Geography in Higher Education* 22(1), 35–48.

Skelton, T. 1989: Women, men and power: gender relations in Montserrat, unpublished Ph.D. thesis, University of Newcastle upon Tyne, UK.

—— 1995: Boom, bye bye: Jamaican ragga and gay resistance. In Bell, D. and Valentine, G. (eds), *Mapping desire: geographies of sexualities.* London: Routledge, 264–83.

—— 1999: Jamaican Yardies on British television: dominant respresentations, spaces for resistance? In Sharp, J., Routledge, P., Philo, C. and Paddison, R. (eds), *Entanglements of power.* London: Taylor & Francis, 182–203.

Smith, M. G. 1963: *West Indian family structure.* Seattle: University of Washington Press.

Smith, R. T. 1988: *Kinship and class in the West Indies.* Cambridge: Cambridge University Press.

Staeheli, L. and Lawson, V. 1994: A discussion of 'Women in the Field': the politics of feminist fieldwork. *Professional Geographer* 46(1), 96–102.

Wolf, D. 1996: Situating feminist dilemmas in fieldwork. In Wolf, D. (ed.), *Feminist dilemmas in fieldwork.* Boulder, CO: Westview Press, 1–55.

# 7

# 'Insiders' and/or 'outsiders': positionality, theory and praxis

## Robina Mohammad

## Introduction

In the third year of my undergraduate degree I went to see a careers officer to talk through my future options. She watched me as I explained the subject of my dissertation thesis, on young Pakistani women, 'othering' and 'otherness'. After listening to me and 'reading' my colour, her advice to me was to embrace the academy in order to research 'Other', Pakistanis like myself. 'A capable person like yourself', she told me, 'can do most things, but there are some things only *you* (as a Pakistani) can do'. In the same year I applied for a post, researching adult education needs in the local Pakistani community. I attended the interview wearing a shortish skirt and blouse, the sort of clothes that I had always worn for job interviews. I got the job, I was told, despite my dress. I later realized I had been expected to dress as, and perform the role of, a Pakistani for the interview. I was told that a choice had been made between myself and another applicant, a young Pakistani woman who had recently arrived from Pakistan. She had apparently fitted the part better than me. Her limited knowledge of English, however, necessary for reporting the findings of the research, meant that they had to settle for me instead.

These episodes highlight two points at which I was confronted with the notion of positionality and the 'insider'/'outsider' boundary with regard to research methodology and knowledge production. 'Insider'/'outsider' refers to the boundary marking an inside from an outside, a boundary that is seen to circumscribe identity, social position and belonging and as such marks those who do not belong and hence are excluded. I was seen to be marked by my colour, which set me apart from 'white' society while positioning me elsewhere, as part of an Asian 'community'. In this sense I was seen to be an 'insider', as someone who was from and hence 'belonged' to the local Pakistani 'community'. This belonging was seen to endow me with a superior, almost organic knowledge of the 'community' not accessible to 'outsiders' – for example, white people. In this chapter I draw on my experiences of this work with the Pakistani 'community' to look at why social positioning and the notion of 'insider'/'outsider' has become relevant to social research and the production of knowledge. I examine why this has not always been so and finally I consider how useful the concept of 'insider'/'outsider' is for ethical fieldwork and as a means of authenticating knowledge (see Fig. 7.1).

**Fig. 7.1** *This is simultaneously a picture of a young girl and an older woman. It is easy to identify one woman, but to see the other requires a shift in perspective. In this way this picture cleverly shows the way in which sight is partial and depends on the perspective of the viewer that frames the research field*

From F. R. Bradbury (ed.), Words and Numbers (Edinburgh University Press, 1969)

## 'Them'

In the 1980s, ideas that ethnic and cultural differences would and should disappear through assimilation into hegemonic cultures began to be replaced by the notion of 'multiculturalism'. This referred to the idea that different cultures[1] should coexist alongside one another without assimilation, emphasizing the need to respect and valorize differences rather than assimilate them. Interestingly this occurred at the same time as a growing emphasis on a collective identity at the national scale, seen, for example, in Britain by the putting in place of a national curriculum (Judd 1989) to facilitate the development of this identity (Gilroy 1987), a boundary distinguishing those who belonged in the nation and were 'insiders' from those who did not and were hence 'outsiders'.

In the academy post-colonial and feminist critiques were demanding the recognition and acknowledgement of social differences of gender, race and sexuality as well as class. These critiques aimed to understand the power relations underpinning these differences and to develop means to include those who had been excluded in these different ways. They linked multiple exclusions to the andro- and ethnocentricity of hegemonic knowledges. In this way they problematized Western, modernist methodologies and epistemologies (what we know and how we know it) as part of their struggles for representation. The work of Palestinian-American intellectual Edward Said (1978) has been important in drawing attention to the ways in which representations do not merely reflect social reality but are also constitutive of it. He charges Western knowledges of Orientalism.[2] Western representations of the East in disparate fields such as travel writing, anthropology, geography and history, he argues, have had a systematic coherence. These representations have enframed and staged the Orient for Europeans. It is Europe's hegemonic systems of representation whose 'life giving power represents, animates [and] constitutes the otherwise silent ... [unknown and hence] dangerous space beyond its familiar boundaries' (Said 1978: 57). The Orient is represented as alternately passive, mute, irrational, feminized, exoticized, and subordinate for the Western masculine gaze, with the will to control, manipulate and incorporate it (Said 1978, 1993; see also Hovsepian 1992; Parry 1992). In this way the West produces the Orient. It is by positing the Orient as different that the West constitutes its own identity as the 'same'. Said has highlighted that there is a politics of representation. This becomes clearer when one defines the term as both 'a speaking of', which describes or depicts a perceived reality, and a 'speaking for', which is the same as the representation of a particular constituency (Spivak 1988).

Previously excluded groups demanding greater power to represent themselves also sought to ensure that their representations would be accepted as authentic and legitimate (Jay 1994). Discursive and even violent struggles to be heard and seen have 'shaken the claims to representativeness of false universalities ...'(Robbins 1992: 49). Hegemonic knowledges have been underpinned by positivist methodologies. Positivism's insistence and reliance on the values of objectivity, neutrality and reliability (where repeating the study will produce the same results), from the choosing of a topic through to the writing-up of the research, have been the source of its legitimacy. These techniques stress geographical distance, both suggestive of, as well as facilitating, neutrality by offering protection from any bias promoting contamination with regard to the research field and process. A disembodied vision, a view from nowhere and so from everywhere, an all-seeing eye, surveys the research field to perform a god-trick. A god-trick refers to the way in which one may claim to view a scene objectively, from high above, enabling one to *claim* to see everything and through this vision *know* everything. Knowledge is then posited as *the* truth, as grand, totalizing, theories implying universal applicability.

Feminists have shown that these truth claims are not necessarily universal, while objectivity and neutrality are myths. All knowledges, they insist, are embedded, situated, specific and hence partial, with an inevitable bias (Haraway 1988; Bhavnani 1994; Bhavnani and Haraway 1994). The researcher is part of the social world; he or she is like 'Jonah ... inside the whale' (Smith 1987: 142), making objectivity and neutrality impossible. Moreover, the claim to objectivity and neutrality serves only to make invisible the biases and subjectivity of the information that is collected and coded as knowledge.

Because researchers have questioned the legitimacy of Western and androcentric knowledges, or truth claims, the legitimacy of all knowledges has been called into question, opening up the need for alternative ways to authenticate knowledge through more ethical methods of knowledge production. The question that has emerged is put simply as who has the right to speak on behalf of whom? And with what authority? (Gregory 1994; Jay 1994; Radcliffe 1994.)

## Professionalism as domination

If Modernist knowledges operated by negating the power relations underpinning them, by obscuring their partiality and situatedness, then a more ethical approach would suggest making the invisible visible or the unconscious conscious, to reveal the situatedness, the enframed and staged nature of knowledges. This would help avoid invalid generalizations. It could be achieved by making visible the exact nature of the biases of the researcher through a self-reflexive understanding of the social locatedness of the researcher and fieldwork relations (McDowell 1992; Gilbert 1994; Mattingly and Falconer-Al-Hindi 1995). This in turn would be the source of its legitimacy and authority. Reflexivity for this purpose must take the form of a particular spatiality, a Janus-like vision that looks inwards at the researcher and outwards towards the researcher's relationship with the researched in the field. Described as a self-conscious, self-critical gaze (Moss 1995), self-reflexivity is seen as a means of discovery of the self (England 1994) that enables a self-awareness of one's positioning in the field (Katz 1994). Rose (1997: 309) has pointed out how this researcher self becomes posited as 'a transparently knowable agent ... [which] looks outward, to understand its place in the world, to chart its position in the arenas of knowledge production, to see its own place in the relations of power'. Concerns of ethical research and legitimacy of knowledge that place an emphasis on uncovering inequalities and exclusions lead to the complexity of power relations made 'into a visible and clearly ordered space' (Rose 1997: 310). The position of those being researched is always defined vis-à-vis the researcher. Power is seen to be distributed unevenly between the two sides. It is often perceived to be held by the researcher (Nast 1994) at the expense of the researched. The relationship of the researcher to the researched, positioned in this way in a landscape of power, is mapped in terms of either 'difference articulated through an objectifying distance; or as a relationship of sameness, understood as the researcher and researched being in the same position, same social location' (Rose 1997: 313). In view of the conscious necessity to avoid objectivity, sameness based on experience becomes linked to authenticity and moral authority to personal history (Harvey 1993). Experiential 'sameness' is used to provide moral authority to an account on the basis that this sameness endows the researcher with greater understanding of the researched's reality.

In this way privilege has become identified with taintedness and oppression with virtue (Bondi 1993). Brah (1992: 135–6) highlights this tendency in the women's movement with the emergence of identity politics:

> *The mere act of naming one self as a member of an oppressed group ... [is] assumed to vest one with moral authority ... the more oppressions a woman could list the greater her claims to occupy a higher moral ground. Assertions about authenticity of personal experience could be presented as if they were an unproblematic guide to an understanding of the processes of subordination and domination.*

Note, for example, the need for Gregory (1994) to defer to bell hooks to explain *the* African-American experience. Barnett (1995) suggests that Gregory's anxiety of being charged with academic neo-imperialism appears to be far outweighed by his fear of paralysis of being 'condemned to speak only of what one knows by way of one's own experience' (Gregory 1994: 205). These anxieties are not restricted to privileged men of the academy. White South African feminist Jenny Robinson (1994: 198) talks of feeling silenced by Black South African feminists, who have accused white feminists of 'never studying white women and of constantly appropriating the experiences of black women for their own personal gain'. Kim England (1994: 84) questions her right to study lesbian communities in Canada, because they may 'perhaps [be] the ultimate "other" given that lesbians are not male, heterosexual, not always middle-class and often not white? ... I worried that I might be, albeit unintentionally, colonising lesbians in some kind of academic neoimperialism.'

With legitimacy, being rooted in experience and experiential knowledge posited as a means of validating truth, the 'Other's' experience of alienation and suffering threatens to become transformed into a profitable commodity (Robbins 1992, 1993; Varadharajan 1992). As 'authentic inhabitants of the margin' (Spivak 1990: 224), 'we' are considered to be positioned legitimately to speak on behalf of 'our' respective groups, 'constituting and adjudicating both the limits of knowledge about the "Other" and the graphematic space the latter occupies/inscribes' (Spivak 1992: 175). This in turn facilitates the upward mobility of 'Other' intellectuals.

In a parallel way, my Pakistani ethnicity was seen to provide me with the authority to research a local Pakistani 'community', as the two anecdotes in the opening paragraph show. I have talked about the way in which I was expected to make myself an authentic Pakistani by dressing appropriately and playing the part. I will highlight some of the ways in which the 'insider'/'outsider' boundary mediated the research process (through the discourses that the researched 'community' and I had both participated in at one time or another in different ways). This will lead me to into a discussion of the limits of positionality as a strategy for authenticating knowledge.

## The Pakistani 'community'

The Pakistani 'community'[3] in this urban locality in the south of England, where I worked, is relatively small. The 1991 census suggested that the Pakistani population numbered around 3000. This population is predominantly working class. There is a high level of unemployment, which in 1991 stood at 20 per cent for men (the highest of all groups) and 30 per cent for women. The majority originate from rural areas in the region of Mirpur in Northern Pakistan.

I was involved in four research projects. These combined techniques of participant observation, semi-structured questionnaires, group interviews and in-depth interviews. For the first of these projects I researched the issue of 'otherness' and identity as part of my undergraduate dissertation. I met a number of young women, for four in-depth interviews (of one–two hours) between May 1994 and January 1995. During the summer of 1994 I was also employed by the Workers Educational Association (WEA) on a project concerned with assessing the adult education requirements of the local Pakistani 'community'. This research was based on questionnaires and group interviews. In 1995 I was also employed (in conjunction with two white colleagues[4]) on a two-part project to be carried out in the summer of 1995 and 1996, looking at Pakistani women's participation in the labour market. This research was actually requested by the 'community' itself. The first part of this research was based on twenty-five in-depth interviews with women under 25 years of age and the second part with twenty-five similar interviews with women over 25 years. I also participated at key community sites such as the Pakistani Community Centre (PCC), the Asian Women's Mother and Toddler Group (AWMTG) and the Muslim Women's Association (MWA), talking to women (and men), in order to gain some understanding of the circulation of discourses (what was being said and felt about general issues) within the 'community'. The participants for in-depth interviews were selected through the snowball method, where initial respondents recruited from 'community' sites were asked to introduce other respondents.

## And 'us'

It is important to note at this point that what we recognize as a collective Pakistani Muslim identity is actively produced through shared rememberings – the telling of stories, in which women and men both participate (cf. Yuval Davies 1997). Social, cultural and religious mythologies are the central themes of these narratives and practices. These myths are given life by histories, by shared experiences of racisms and imperialisms and by opposition to the West. The media, popular culture and literatures all instruct everyday rituals that define 'us', distinguishing 'us' from 'them'. This border is celebrated and reaffirmed through statements such as:

> we are different from them. My husband and I have sought to keep our children apart by not allowing them to attend birthday parties [of white children] etc., so they would understand and recognize that we are not like them and can never mix with them.
>
> (Mrs Butt,[5] taped interview)

and

> *English people are different from us. They don't understand us. They think because they*
> *are white they are better than us. They aren't.   (Shela, taped interview).*

The political commitment to multiculturalism also plays a role in these processes, as it makes accomplices of 'Others' in the policing of the border. One respondent noted how in school:

> *If they [i.e. the teachers] find some of the Asian girls doing things they shouldn't be doing*
> *then, because the teachers are so aware of Islamic culture and the Hindu culture and all*
> *the other multi-cultures, they do inform the parents and the parents are aware of that ...*
> *(Leila, taped interview)[6]*

In this way an 'us' is always in the process of being constructed and filled with meaning, offering a range of subject positions for Pakistanis that must be negotiated, contested or resisted.

Before I begin discussing my interactions with the Pakistani community, I should say something about my own positions. I see myself as a British, Pakistani Muslim (by birth), but non-practising and non-believing, a little Marxist, somewhat feminist, of middle working-class origins. I am divorced and in a relationship with a white man. Most of these details, however, were shared with only a few of the younger respondents, because I felt they would be non-judgemental about these facts. These were also people who shared similar information about themselves with me and so I felt safe to share information with them. The sharing of personal information is part of everyday interaction, but in a research setting it becomes a conscious act. It becomes necessary to consider whether to share information and if so what information to share, both for the researcher and for the researched. In this case what I shared with my respondents also changed over time. As we built up an element of trust, we were able to share a little more. In the early stages, like them, I too was concerned about confiding too much information that might get known in the 'community'. I was concerned that, if certain information about me did become common knowledge, it would attract moral disapproval from the 'community' in a way that it would not for an English/white woman researcher. These concerns were confirmed by the way in which the 'community' saw me as one of 'them' in all the ways that mattered – that is, as a Pakistani woman and a practising Muslim. My positioning by them as an 'insider' in this way often led to an erasure of the personal/professional boundary. As the researched closed the distance between us, I was expected to attend the Friday prayers at the MWA.

In my interactions with Pakistanis in the field, I was always acutely conscious of the fact that I had committed the ultimate sin of straying beyond the boundary of the 'group' (that is seen solely in terms of ethnicity, irrespective of differences of geographical and social location) in having a (non-marital) relationship with a white man. This highlights the way in which I became reinterpellated[7] by a familiar discourse whose invitation I had stubbornly attempted to resist, not always successfully. During my final period of fieldwork I became pregnant. Although my work involved talking to women with small babies and toddlers at the Pakistani women's mother and toddler group and under different circumstances I might have shared this fact, I decided against revealing it to avoid drawing attention to my marital reality. My divorced status had been concealed from the 'community'. It was agreed that I would be presented to the group as Mrs Mohammad to give me a more respectable identity, which would inspire confidence in Pakistani male community leaders to allow me to interact with young women for the research. This may be regarded as unethical; I would argue, however, that in interactive research contexts the researcher and 'community' being researched are performing roles. Neither party really knows the 'truth' of these roles. We learn snippets of information about each other, but neither party can see the whole play at work nor understand the total reality of one self let alone others. In this way the respondents performed a given role or set of roles to me and I to them. For some of them I was the divorced, independent, 'Westernized' woman they aspired to be and so a positive thing; for others this was negative. For yet others, I was a Muslim who could join them for prayers at the Mosque, an identity that I did nothing to refute. In this way, although I was presented to the 'group' as a married woman, the 'group' in this case was a few powerful patriarchs from the Pakistani 'community' who did not necessarily reflect the concerns of the group as a whole. For many of the women I encountered and engaged with my status was a positive thing, while for others it did not seem to arise as an issue.

Despite the play of different identities, on the whole the researched 'community' did endow me with 'insider' status. They tended to take my knowledge regarding Pakistani 'culture' for granted. I would often hear remarks such as 'we shouldn't really say this since it is disloyal [to the group] but since you are one of us it is all right', or 'I can tell you because you'll know what I'm talking about' (by both men and women). These comments were followed by some sharing of gossip about the group or personal experience that otherwise, it was suggested, would have been withheld or presented differently. I found that I also took some knowledge for granted so was not aware of highlighting certain issues that cropped up in the research until others in the academy pointed out specific instances when this happened. This was the case, for example, with respect to the degree of differences in the geographical mobility of young Pakistani men and women, something that I simply took for granted. The world of young, working-class, Pakistani men was much bigger than for young women. The freedom to roam about town means that they are able to indulge in activities that are prohibited more easily than young women, which impacted quite significantly on women's life experiences. In this way the research 'findings' can be seen to be 'produced' out of a set of interactions and there will be differences in what even different researchers of the same colour, gender and/or class may find or give significance to. Even a researcher returning to the same field may find and give significance to a different set of data and a different research story.

Working with this 'community', which was in some ways for me, as well as for them, my own, brought into sharp relief the latent stereotypes and taken-for-granted assumptions that, not altogether consciously, I had held. Years earlier, as a young, new mother, dressed in Punjabi dress and hair scraped back, I had been annoyed by the assumptions of my health visitor, who spoke to me very slowly, taking care to pronounce her words clearly to help me understand her. The shock on her face as I responded to her in my Geordie accent showed that she too was aware of these assumptions, although nothing was verbalized. During the course of this research I found myself making similar assumptions. In my meeting with a young woman at the PCC-based mother and toddler group, I pondered whether I should talk to her in Punjabi, because I was not sure that she would understand English. I later found out that she was a law graduate and realized that I had presumed from her appearance that she was a much older, semi-literate woman from rural Pakistan. She was rather rounded, wore quite a lot of jewellery, and her hair again was simply tied back. In my experience of Pakistani women, the style of clothing and hair is a significant indicator of education and social status. Women from rural backgrounds are often less educated and tend to wear more jewellery on an everyday basis, while urban women tend to wear a lot of jewellery only on special occasions – hence my assumptions. This incident again made me reflect on the fact that there is no singular Pakistani identity. Just as I was imputing a singular identity to these women, others have imputed to me a singular identity, which is seen to give me a privileged access to knowledge about Pakistanis.

This positioning of the researcher as an 'insider', however, can also have the effect of making the researcher appear too close for comfort, making people wary of sharing information. On the project concerned with the experiences of women in the labour market, I worked with two white and therefore 'Other' academics, from the point of view of the researched. We found that the members of the 'community', who themselves had called for this research and specified the necessity of a Pakistani, Muslim researcher, subsequently found some difficulties with this. This group of women at the PCC, having decided on the type of collective self they wanted to present, both to the researchers and to the agencies commissioning the research, sought to shift attention away from issues such as arranged marriages in order to avoid stepping into what they presumed to be 'our' stereotypes of them (perhaps as oppressed women?). They did this by denying to the white academics that such practices existed in this community and moreover by prohibiting the questioning of young girls on issues related to marriage and motherhood.

My involvement in the project occurred a few weeks later. Having established these 'truths' with the white researchers and having forgotten about my involvement, these women appeared 'caught out' by my arrival on the scene. Their attention now also became focused on concerns of confidentiality. Assuming that I was part of the 'community' (perhaps encouraged by my Punjabi dress), they worried that my privileged access to information about their daughters, who were to be the initial focus of the study, could easily leak out, inadvertently perhaps, and spread throughout the small community. In this situation I had to assert my 'outsider' position in terms that they could relate to – for example, emphasizing my geographical distance, explaining that I had at one time belonged to a 'community' that, while being very similar in cultural terms, was geographically speaking very distant from this 'community'.

This 'insider'/'outsider' border also mediated both mine and the researched's performances in the situation of the interview. The respondents performed particular identities to me, negotiating different, simultaneously illusory and real boundaries through contestation and resistances from those they regarded as the 'Other' (white), the 'Other within' (the Pakistani male) and me (seen variously as 'Same', 'Other' and simultaneously 'Same' and 'Other').

One interviewee, noting my 'Westernised' appearance and armed with the knowledge that I was a divorcee with two children living independently and studying for a degree, felt able to perform for me the role of an 'English' girl, most immediately through visual markers such as dress. For her, this was a means of resisting the category of the 'virginal', 'pure' Muslim woman constructed by the identity politics discourse of the 'group'. Part of this strategy involved making me audience to her sexual history, complete with photographs of her different boyfriends, geographically dispersed to minimize the risk of discovery by her parents and the 'community'. She set the scene with the revelation that 'playing the perfect daughter in the home, by dressing in Punjabi dress and covering my hair, [buys me trust and freedom] to be the imperfect daughter outside it'. She confided in me how, even as a child, she had hated everything Asian:

> I always hung out with English girls; when we had non-uniform days all the other Pakistani girls would dress in Punjabi dress but I just wore my uniform because I felt it was more me. Today I dress in Asian clothes only at home and at weddings, when I feel pretty and sexy in them and the guys also like them. I watch Indian movies and keep up with the fashions, but I still feel more myself in English clothes.    (Sharon,[8] taped interview)

It is clear that she is interpellated by the group identity discourse, as, despite her resistances to the subject position offered to her, she remains within the binary of good Muslim girl and bad English girl – that is, the 'insider' and 'outsider' positions offered by the discourse. Either she is offered the position of being 'inside' the group, for which she must play the role of a good, Muslim girl; or she may exclude herself by resisting the role of good, Muslim girl to play the bad, English girl. In reality she plays both roles in different spheres and is unable or unwilling to seek alternatives to these identities. This reveals the impact of the discourse on the psyche that undermines the development of other identities. Towards the end of the research period she confided in me that

> I only agreed to speak to you because of what I had heard about you being divorced and
> living on your own. When I met you and your children I thought that they were very
> English, I thought she's alright and decided to tell you everything. Had you dressed in
> Punjabi dress and looked Pakistani ... I would have refused to speak to you.    (Sharon,
> taped interview)

My status in terms of the 'insider'/'outsider' border was, it seems, highly ambiguous. For Sharon, as for Shela, my authenticity was uncertain. On my initial meeting with Shela, the boundary between 'Other' and 'Same' stood between us. Once she had declared to me that 'I value our tradition, I am proud to wear our traditional costume', my appearance in jeans seemed to signify a betrayal of my commitment to 'our' traditions. They provided the evidence for her of the 'cultural gulf' between us. Her acceptance of my authenticity, however, was visible on her face when she discovered that I was both fluent and literate in Urdu (the national language of Pakistan); it was clear that I now qualified as a Pakistani.

Identity politics discourse did not interpellate young women in any simple or automatic way. If respondents were uncertain regarding my authenticity, they were not altogether secure about their own positioning. One respondent remarked on her sense of 'confusion' about where she belonged: 'as I have been living here in the UK ... and have grown up here, I do not feel that I will fit in if I was to go back home [to] Pakistan, as I feel British. I do know, however, that [the] UK is not our homeland' (Tara, taped interview). She then turns to language in an attempt to reconcile her sense of belonging more in the UK with the knowledge that the UK is not her homeland: 'as I have been brought up here, I always speak English. I feel my first language is English, as I always think in English' (Tara, taped interview). A third respondent's negotiation of this position is exemplified by her reflection that she is a 'British Pakistani with Muslim beliefs ... [who] has what some people might call English ways and thoughts' (Marina, taped interview) – which were declared to be an openness to the idea of premarital sex and an insistence on having a career. Increasingly, young Pakistani Muslims, born and brought up in the West, are challenging discourses that position them in opposition to the West.

As young Asian Muslims identify with global Islam, they are able to see the way in which Islams have been shaped within different national/cultural contexts so that they are able to pick and choose between different ways of practising Islam. Valley and Brown (1995b) note, for example, how young Muslim women are increasingly moving away from their parents' interpretation of Islam towards an Islam in which they see no contradiction between Islam and feminism (Dwyer 1999), a discourse posited as Western by their parents' interpretations (Mohammad 1999). The opposition of Islam to the West is called into question by young British Muslims such as Jamil Ali. Ali notes how 'there was in Islam a tradition which painted the devil as being in the West, that won't do for us, we are the West' (Ali, cited by Valley and Brown 1995b: 4).

## Conclusion

The focus on self-reflexivity and positionality translates into the 'insider'/'outsider' border. This border then acts as container for difference and similarity that are posited as homogeneous categories. In this way the 'Other' intellectual becomes endowed with a privileged capacity to understand the experiences of those similarly excluded. The form of that exclusion, however, can never be the same. The WEA's search for a candidate who was more 'authentically' Pakistani, an unequivocal 'insider' for the job, was frustrated by the fact that her very authenticity was derived from her lack of the skills necessary for the job.

The border is underpinned by 'a conception of identity as something to be acknowledged or uncovered rather than constructed, something fixed rather than changing' (Bondi 1993: 93). It is represented as something complete rather than always in the process of becoming (Hall 1993). A recognition of difference has led to the practice of hyphenating pre-given identities so that 'multiple oppressions ... [become] regarded not in terms of their patterns of articulation/interconnections but rather as separate elements that ... [can] be added in linear fashion' (Brah 1992: 135). Thus, for example, black-working-class-lesbian, which suggests that affliction by different oppressions *multiplies* the amount of suffering (see England 1994). I would argue, however, that, rather than an adding-on, there is an intersection of exclusionary subject positions that qualitatively change the nature and experience of oppression (see Mohammad 1999).

If one takes into account that identity and 'subjectivity ... [are] linguistically and discursively constructed and displaced across the range of discourses in which ... concrete individuals [may freely] participate ... [and that this] matrix of subject-positions ... may be inconsistent or even in contradiction with one another' (Belsey 1980: 61), then difference cannot be fixed or containable within oppositional categories of 'same', 'insider', 'outsider'. These categories, like pronouns, are empty and shifting, filled with meaning contextually by those who 'wear' them. In this way I can be both an 'insider' and an 'outsider', and both 'insider' and 'outsider' in the same place and/or time (Benveniste 1971).

If identity then is constituted in discourse with multiple webs of intersecting discourses, how far can one be self-reflexive in the ways advocated by feminists? Is it not the case that to be so fully self-conscious and self-aware of the field and one's position in it suggests a similar god(dess?)-trick against which self-reflexivity has been recommended as a means of making visible the partial nature of one's knowledge (Rose 1997).

The tactics to make power relations visible in the production of knowledge often lead to reconfirming the power of the powerful and the oppression of those deemed to be powerless. Said's critique, for example, confirms Western power and the truth of the West's representations of the East. In the same way, ethical researchers who survey the field, self-conscious of power that they appear to hold, and the need for 'shifting a lot of power over to the researched' (England 1994: 82), or the 'struggle to distribute power more evenly'(Farrow et al. 1995: 71), simply reconfirm it. It then follows that their view of the researched must also be total, that there could not be an aspect of the field and the researched that is not transparently visible to them (Katz 1996). Varadharajan detects the traces of a self-consciousness about one's own power even in the concern to refrain from 'speaking on behalf of those whom [one] can never "know" '. She argues that this 'presumes that having spoken [one] would have said it all and that the other will be moved neither to challenge nor to supplement ... [this] humility has all the trappings of a gracious and patronising self-effacement ...' (1992: 3). In this way epistemological questions raised by the 'crisis of representation' translate into the dilemma: 'of finding oneself unsure of the grounds of one's own representational utterances, manages to keep the focus of attention on the anxieties of a certain privileged form of subjectivity, now reduced to narrating the tragic inevitability of its own dissembling' (Barnett 1995: 428).

In the field, I have been only too aware of the limits to my ability to control interactions with my respondents. While I could direct an interview, I could not make any one answer my questions or quieten someone who wanted to talk about a topic not on my list. I had a rough research agenda to follow, but what was created out of the process of fieldwork has never been entirely what could have been predicted – a lot of loose ends and tangled webs that then need to be discarded or woven into coherence if possible with some consciousness and some authority. In this way the knowledges produced are always versions. They represent one out of other possible truths. The male figures in the PCC, for example, would not see my accounts as truthful. The women at the PCC concerned with the mystification of the arranged-marriage issue would not accept my version as entirely truthful because I have talked about young women's strategies for negotiating these issues. The heterogeneity of the researched makes impossible the notion of a singular truth. The researched had their own interests, agendas and ideas of what they hoped to gain from the research that were tied up with the presentation of a particular collective self to 'us'. Many of the older women felt that they were representing Muslim women and the Muslim world. The majority of younger women, however, saw their involvement in the research as a means to increase awareness of the everyday oppressions they experienced, whether these were produced by patriarchies, racisms or both. They also seemed to take the opportunity to 'wear' and make visible other identities that they were not always able to do in other contexts.

Given the complexity of researcher's and researched's identities and the dynamism of the research field, truth claims can be grounded only in a real recognition of the limitations of vision and knowledge and the existence of multiple truths. This is not to say that all truths have the same value. Their value must be ascertained from their potential political effects. Instead of asking the question 'is this true?', it is more useful to ask 'which truth?' or 'whose truth?' is being told. Whose interests are being served by a particular representation (Donald and Rattansi 1992)? In this way it is possible and indeed necessary to affirm the legitimacy of some representations over others. It is, therefore, the researcher's/author's responsibility to choose and affirm his or her political commitment through the types of representations he or she seeks to make, irrespective of his or her social position, colour or gender. The author, however, must also recognize that he or she has no absolute control over language, meaning and representation; that this control is shared with readers, who may change and/or supplement meanings through the act of reading.

## Notes

1. These cultures were often defined in relation to hegemonic cultures and posited as homogeneous, pure and fixed.
2. Said (1978) appropriated the term from the academic discipline devoted to the study of the East. He expanded the term to refer to 'a style of thought based on an ontological and epistemological distinction made between the "Orient" and ... the "Occident" ' (Said 1978: 2).
3. I use the term 'community' to highlight the fact that it is heterogeneous and fragmented but this diversity and disunity are negated by a discursively produced suggestion of a collective identity, which I have discussed elsewhere (Mohammad 1999).
4. Sophie Bowlby and Sally Lloyd-Evans.
5. These are pseudonyms that I have given the interviewees; older women are referred to by family names and younger women by first names.
6. This commitment can be seen in different arenas and for different purposes. For the groups of the left, it is a move to recognize the diversity of Britain's population. For leaders of ethnic groups, it facilitates the purchase of financial aid/support. It also enables such leaders to establish their groups' right to be separate and different. In this way they seek to normalize difference but are also simultaneously reconfirming it. This can be seen in the demand for Muslim schools for Muslim girls. Muslims, just like Catholics, it is argued, have the right to educate their children in their own religion and so maintain their difference.
7. Interpellation refers to the way in which we are constituted as subjects in language through the pronouns of I, we, us, them. These pronouns literally draw us into positions, for example, of 'us' or 'we'. It is by wearing the I through enunciation that the speaker posits him or herself as the subject of the sentence (Belsey 1980). In this way, when some one shouts 'Oi you over there' and you look up, you have been interpellated as the 'you' that the speaker is referring to. In a similar way, when I was confronted by a discourse that spoke of Pakistani Muslim women, virginal or married, obedient and home loving, a discourse that spoke of me as that Pakistani woman, I slipped almost subconsciously into this role and then felt guilty because I was not those things. Interpellation is therefore the positioning of an individual into a particular role in a given narrative. It is a positioning that requires the active involvement of the individual his or herself, in the same way that when watching a movie we may identify ourselves with a particular character and view the movie from the perspective of that character. The individual can slip into the role created by the discourse or resist it.
8. This name was chosen to reflect this young woman's vision of her own position within the Pakistani 'community'.

## Acknowledgements

I would like to thank the editors for their help in reworking this chapter and James Sidaway for his comments.

## Key references

Haraway, D. 1988: Situated knowledges: the science question in feminism and the privilege of partial perspective. *Feminist Studies* 14, 575–99.

Jay, G. S. 1994: Knowledge, power and the struggle for representation. *College English* 56, 9–27.

McDowell, L. 1992: Doing gender: feminism, feminists and research methods in human geography. *Transactions of the Institute of British Geographers* NS 17(4), 399–416.

Robbins, B. 1992: The East as a career. In Sprinkler, M. (ed.), *Edward Said: a critical reader.* Oxford: Blackwell, 48–73.

Rose, G. 1997: Situating knowledges: positionality, reflexivities and other tactics. *Progress in Human Geography* 21(3), 305–20.

Spivak, G. C. 1990: Post-structuralism, marginality, postcoloniality and value. In Collier, P. and Geyer-Ryan, H. (eds), *Literary theory today.* Cambridge: Polity, 219–44.

## References

Althusser, L. 1971: *Lenin and philosophy and other essays,* trans. Ben Brewster. London: New Left Books.

Barnett, C. 1995: Why theory? *Economic Geography* 71(4), 427–35.

Belsey, C. 1980: *Critical practice.* London: Routledge.

Benveniste, E. 1971: *Problems in general linguistics.* Miami: University of Miami Press.

Bhavnani, K. K. 1994: Tracing the contours: feminist research and feminist objectivity. In Afshar, H. and Maynard, M. (eds.), *The dynamics of 'race' and gender: some feminist interventions.* London: Taylor & Francis, 26–40.

—— and Haraway, D. 1994: Shifting the subject: a conversation between Kum-Kum Bhavnani and Donny Haraway on 12 April, 1993, Santa Cruz, California. *Feminism & Psychology* 4(1), 19–39.

Bondi, L. 1993: Locating identity politics. In Keith, M. and Pile, P. (eds), *Place and the politics of identity.* London: Routledge, 84–103.

Brah, A. 1992: Difference, diversity and differentiation. In Donald, J. and Rattansi, A. (eds), *Race, culture and difference.* London: Sage/Open University, 126–45.

Donald, J. and Rattansi, A. 1992: Introduction. In Donald, J. and Rattansi, A. (eds), *Race, culture and difference.* London: Sage/Open University, 1–8.

Dwyer, C. 1999: Contradictions of community: questions of identity for young British Muslim women. *Environment and Planning A* 31, 53–68.

England, K. V. L. 1994: Getting personal: reflexivity, positionality, and feminist research. *Professional Geographer* 46, 80–9.

Farrow, H., Moss, P. and Shaw, B. 1995: Symposium on feminist participatory research. *Antipode* 27, 71–4.

Gilbert, M. 1994: The politics of location: doing feminist research 'at home'. *Professional Geographer* 46(1), 90–6.

Gilroy, P. 1987: *There ain't no black in the Union Jack: the cultural politics of race and nation.* London: Routledge.

Gregory, D. 1994: *Geographical imaginations.* Oxford: Blackwell.

Hall, S. 1993: Cultural identity and diaspora. In Williams, P. and Chrisman, L. (eds), *Colonial discourse and post-colonial theory: a reader.* New York: Harvester Wheatsheaf, 392–403.

Haraway, D. 1988: Situated knowledges: the science question in feminism and the privilege of partial perspective. *Feminist Studies* 14(3), 575–99.

Harvey, D. 1993: Class relations, social justice and the politics of difference. In Keith, M. and Pile, P. (eds), *Place and the politics of identity.* London: Routledge, 41–66.

Hovsepian, N. 1992: Connections with Palestine. In Sprinkler, M. (ed.), *Edward Said: a critical reader.* Oxford: Blackwell, 5–18.

Jay, G. S. 1994: Knowledge, power, and the struggle for representation. *College English* 56, 9–27.

Judd, J. 1989: Thatcher changes the course of history. *Observer,* 20 Aug., p.1.

Katz, C. 1994: Playing the field: questions of fieldwork in geography. *Professional Geographer* 46(1), 67–72.

—— 1996: The expeditions of conjurors: ethnography, power and pretense. In Wolf, D. (ed.), *Feminist dilemmas in fieldwork.* Boulder, CO., and Oxford: Westview Press, 170–84.

McDowell, L. 1992: Doing gender: feminism, feminists and research methods in human geography. *Transactions of the Institute of British Geographers* NS 17(4), 399–416.

Mattingly, D. and Falconer-Al-Hindi, K. 1995: Should women count? A context for the debate. *Professional Geographer* 47, 27–35.

Mohammad, R. 1999: Marginalisation, Islamism and the production of the 'Other's' 'Other'. *Gender Place and Culture* 6(3), 221–40.

Moss, P. 1995: Embeddedness in practice, numbers in context: the politics of knowing and doing. *Professional Geographer* 47, 442–9.

Nast, H. J. 1994: Opening remarks on 'women in the field'. *Professional Geographer* 46(1), 154–66.

Parry, B. 1992: Overlapping territories and intertwined histories: Edward Said's post-colonial cosmopolitanism. In Sprinkler, M. (ed.), *Edward Said: a critical reader.* Oxford: Blackwell, 19–47.

Radcliffe, S. A. 1994: (Representing) post-colonial women: authority, difference and feminisms. *Area* 26(1), 25–32.

Robbins, B. 1992: The East as a career. In Sprinkler, M. (ed.), *Edward Said: a critical reader.* Oxford: Blackwell, 48–73.

—— 1993: *Secular vocations: intellectuals, professionalism, culture.* London: Verso.

Robinson, J. 1994: White women representing/researching 'Others': from anti-apartheid to post-colonialism? In Blunt, A. and Rose, G. (eds.), *Writing women and space: colonial and postcolonial geographies.* New York: Guildford Press, 197–226.

Rose, G. 1997: Situating knowledges: positionality, reflexivities and other tactics. *Progress in Human Geography* 21(3), 305–20.

Said, E. 1978: *Orientalism.* New York: Vintage.

—— 1993: *Culture and imperialism.* London: Vintage.

Smith, D. 1987: *The everyday world as problematic: a feminist sociology.* Boston: Northeastern University Press.

Spivak, G. C. 1988: Can the subaltern speak? Speculations on widow sacrifice. In Nelson, C. and Grossberg, L. (eds), *Marxism and the interpretation of culture.* London: Macmillan, 271–313.

—— 1990: Post-structuralism, marginality, postcoloniality and value. In Collier, P. and Geyer-Ryan, H. (eds), *Literary theory today.* Cambridge: Polity Press, 219–44.

—— 1992: Theory in the margin: Coetzee's Foe reading Defoe's Crusoe/Roxana. In Arac, J. and Johnson, B. (eds), *Consequences of theory.* Baltimore: Johns Hopkins University Press, 154–81.

Stanley, L. and Wise, S. 1993: *Breaking out again: feminist ontology and epistemology.* London: Routledge.

Valley, P. and Brown, A. 1995a: Pride and prejudice. *Independent,* 5 Dec., 2–5.

—— —— 1995b: The best place to be a Muslim. *Independent,* 6 Dec., 2–4.

Varadharajan, A. 1992: The post-colonial intellectual and the question of Otherness. Unpublished paper presented at Gender and Colonialism Conference (Ireland), 1–16.

Yuval Davies, N. 1997: *Gender and nation.* London: Sage.

# PART 3

## Group discussions

# 8
# The focus-group experience
## Tracey Bedford and Jacquelin Burgess

## Introduction

Most social-scientific techniques rest in dusty obscurity, brushed off and given an airing only within the pages of methodology textbooks as students search for justifications for their dissertation proposals. Not so the focus group. This piece of qualitative methodology has been subject to large amounts of publicity and no little ridicule – 'hocus pocus focus groups' as one cartoon put it a couple of years ago – since it became known that Tony Blair and New Labour relied so heavily on focus groups to help them win the 1997 election in the UK. One consequence of this publicity is that almost anyone bringing two or more people together for any reason whatsoever will call the meeting a 'focus group'. In this chapter, we will discuss the characteristics of the focus group as a research technique, drawing on our recent work for the British Retail Consortium (BRC) on discursive constructions of environmental responsibility by different sectors in the retail commodity chain.

We define the focus group as a one-off meeting of between four and eight individuals who are brought together to discuss a particular topic chosen by the researcher(s) who moderate or structure the discussion. Members of the group may be asked simply to talk to each other or, more commonly, they may be set a series of small tasks, which provide the focus for their discussions. For example, they might be asked to watch a video clip or comment on an advertisement for a specific product. Focus-group meetings normally run for 90–120 minutes. Although we have a preference for using in-depth groups when time and resources allow (as these groups meet repeatedly, developing a trusting relationship, which allows for a more open and honest conversation), we regularly use focus groups as an efficient and interesting way of gaining insight into the ways in which people construct environmental and social issues; share their knowledge, experiences and prejudices; and argue their different points of view.

The methodological leap into group techniques began in the late 1940s, and focus groups have been used in many different social-science disciplines since then. One common use of the technique is to gain understanding of the ways in which members of the public talk about an issue prior to the design of a psychological experiment or a social survey questionnaire (see Lunt and Livingstone 1996). The focus group became part of the market researchers' methodological tool kit in the late 1970s and has since been used routinely as a means of exploring consumers' attitudes and behaviours. In the late

1990s, as we have already noted, the focus group became an important technique to increase public participation in political debate (Lunt and Livingstone 1996; Kitzinger and Barbour 1999). Whilst there are plenty of books ready to tell practitioners how to use focus groups, the full potential and implications of using the focus group remain under-researched. There is relatively little evaluation of the strengths and weaknesses of the technique within geography specifically (see Goss 1996) and the social sciences more generally (Kitzinger and Barbour 1999).

In this chapter, we will draw on our experiences of running focus groups since 1986, concentrating mainly on the conduct of a case study we completed in 1999 as consultants to the British Retail Consortium (BRC). This study is particularly interesting because we used many different techniques to recruit and run the focus groups. Members of the groups included very senior individuals from some of the largest retailing companies in the UK; civil servants from central government departments; and influential campaigners from pressure groups as well as members of the general public. The methodology, as with most research, had its problems and we shall suggest how to avoid some of the pitfalls in this chapter.

## The case study

The BRC, who represent approximately 90 per cent of all retailing companies in Britain, wanted research to provide a discussion paper for a conference on sustainable development that they were planning for the autumn of 1999. Early in 1999 we were asked by the BRC to conduct a case study looking at consumers' perception of the extent to which retailers demonstrated environmental responsibilities in their commercial practices. The BRC were familiar with market research that used questionnaires to determine consumer attitudes. They were aware that there was a large 'value-action' gap between what the consumer said about environmental issues and what concerns they were really motivated by when they went shopping. The BRC, therefore, wanted to use a methodology that allowed the concerns of consumers to be examined in greater depth than that provided by a structured questionnaire. We suggested that, rather than simply asking consumers and retailers about their perceptions of environmental responsibility, we should ask a range of actors throughout the chain of provision and consumption what environmental responsibilities they and other actors undertook. This network approach would allow us to explore who currently took their environmental responsibilities seriously, and where the groups felt environmental responsibility would be best placed in the chain of provision and consumption.

Box 8.1 shows the details of the project in summary. We ran 11 focus groups in total. Four groups were held with *consumers*; four with *retailers*, grouped by size and product sector. This meant the research would concentrate on the opinions of consumers and retailers – the groups of primary interest to the BRC. However, we added three further groups who were felt to have a role to play in ensuring environmental responsibilities were exercised throughout the commodity chain. These groups were *regulators* who

## Box 8.1 Details of the BRC focus groups

| Date of focus group | Sector in the commodity chain | Participants [numbers recruited] |
|---|---|---|
| 24 May 1999 | SME retailers | 4 [5] |
| 25 May 1999 | Grocers | 6 [8] |
| 25 May 1999 | Clothing and mixed retailers | 8 [8] |
| 2 June 1999 | Electrical and DIY retailers | 4 [5] |
| 5 May 1999 | ABC1 males | 10 [11] |
| 5 May 1999 | ABC1 females | 10 [11] |
| 6 May 1999 | C2DE males | 10 [11] |
| 6 May 1999 | C2DE females | 10 [11] |
| 26 May 1999 | Regulators; government depts | 7 [6] |
| 8 June 1999 | NGOs | 3 [7] + 2 interviews |
| 10 Aug. 1999 | Suppliers | 4 [6] |

decide on legal and other standards for the retail industry, *non-governmental organizations* and pressure groups, and individuals representing different *manufacturers/suppliers* to retailers. The groups were recruited and run during the summer of 1999; transcripts were analysed and a report submitted to the BRC at the end of the year (Bedford and Burgess 1999). (For a number of reasons, the conference did not take place.)

## Why use focus groups?

### The benefits

Focus groups place the individual in a group context, where conversations can develop and flourish in what could be considered more commonplace social situations than being interviewed for a questionnaire survey (Lunt and Livingstone 1996). The flow of conversation ensures that there is a dialogue between people, with individuals free to challenge the interpretation or assumptions of other group members. This *dialogic* characteristic of the focus group gives the researcher access to the multiple and transpersonal understandings that characterize social behaviour (Goss 1996: 118). In one-to-one interviews, the individual is usually free to express his or her opinions without challenge – the interviewer may ask for clarification or deepening of a point but will rarely contest what the respondent says. In the group context, people's opinions and beliefs can be questioned and/or amplified by others in the group. In this way, each group creates a unique dynamic as relationships are built up between members on the basis of their dialogic interactions. Some groups are very consensual – with everyone largely agreeing – while others can be very argumentative – with individuals debating energetically among themselves. Because focus groups tend to promote enjoyable discussions (Goss 1996: 117), researchers are able to gain insight into the social, cultural, political, economic and personal dimensions of an issue – its *discourse* – without informants feeling their knowledge is being challenged by a more knowledgeable researcher. For example, in discussing the causes of climate change an individual informant might quickly come to feel ignorant if she kept having to say 'I don't know' in answer to a researcher's questions.

Groups often provide researchers with surprising insights. One of the important benefits of qualitative research is the extent to which researchers, as well as participants, can learn through the process. Conversations take on their own dynamic, and spontaneous group debates can reveal unexpected findings (Stewart and Shamdasani 1990; Holbrook and Jackson 1996), in a way that is less likely to happen in structured interviews or questionnaires. This was particularly helpful in the BRC research in the context of the professional and corporate groups. Asking these groups to discuss their environmental responsibilities produced a rich picture of a variety of environmental initiatives and legislations, interwoven with different views and experiences about the difficulties of implementing new initiatives in different contexts. As we had a very tight period of time in which to conduct the project (four months in total), the focus-group methodology allowed us to obtain much background information that we would otherwise have had to have spend a considerable amount of time hunting out. Indeed, focus groups are often at their most useful when only a limited amount of time can be spent in the field (Burgess 1996).

Focus groups are especially useful when you want to compare the 'world-views' of different sectors or groups of people in an efficient way. Each group is recruited to represent people of similar background, experience or interest (e.g. Macnaghten *et al.* 1995; Burgess 1996) to provide common ground for their discussions. Focus groups are particularly effective at establishing a group identity quickly, and therefore are a good way of testing the attitudes one group displays towards another (Krueger 1994: 45). In terms of research design, the aim is to ensure homogeneity within the group and heterogeneity between them. In the BRC case study, we wanted to explore differences between the sectors – what each group knew of the others' environmental responsibilities and their attitudes towards them. What did suppliers, retailers, regulators, middle-class and working-class consumers, and pressure groups feel to be their specific responsibilities for reducing the impacts of human activity on environmental systems, and which sectors did they feel were being remiss in exercising their responsibilities?

### Some problems

Having outlined these advantages, it is important to acknowledge some of the difficulties, especially for younger researchers and undergraduates doing dissertations using a technique that is dependent upon group dynamics and relatively free-flowing conversation. Groups can discuss with enthusiasm a topic in which, on an individual level, the members have little interest (Lunt and Livingstone 1996). At the other extreme, it is sometimes impossible to promote a topic of conversation because the group has *no* interest in discussing it. Nor is it possible to be sure prior to facilitating the group whether the group will move towards consensus or dissent, as this is dependent upon both the topic and the individual characteristics of the group members.

Unlike in-depth groups, it is difficult to screen closely the potential members of focus groups (see Harrison *et al.* 1996). Researchers are vulnerable to individuals who may, for example, claim authority to represent the views of others when they do not; or who find the presence of a 'captive' audience just irresistible and never stop talking. In one of the BRC retailer groups, a particularly dominant individual insisted that, in his retail sector, the major barrier to improving environmental performance was the irresponsibility of customers. After the group had dispersed, another group member came back to tell us that the vociferous individual was not representing that particular retail sector fairly. In academic research, the problem of representation is overcome by running as many groups as it takes to ensure that the stories used have all been heard many times (Lunt and Livingstone 1996). But this is very demanding of both time and financial resources. Normally, the way to deal with the problem is to be cautious in writing up the findings – always express findings as aspects of the case study, rather than a representative study of the entire system of provision and consumption.

## Recruitment of focus-group members

The techniques used to recruit focus groups are as important as the techniques used to run them. Possibly the most important decision to make is about the composition of the groups in the research design (Stewart and Shamdasani 1990: 51). The choice will be justified by the theoretical-conceptual issues underpinning the research (see e.g. Burgess 1996). Additionally, age, gender, socio-economic class, religion and race are all likely to affect the ways that individuals interact with each other (see Stewart and Shamdasani 1990: 36–46 for a fuller discussion). This is as much an opportunity as a problem, and the task for researchers is to decide whether they would wish to recruit groups that are representative of a community or organizational context, or whether it is more appropriate to have groups with common demographic characteristics in order to be sensitive to demographic impacts. For example, in the BRC research we ran four groups of consumers, differentiated both by gender and by socio-economic class. The decision was made on the basis that people consume differently depending on gender (Pringle 1987; de Vault 1991; Dowling 1993; Lury 1996; Campbell 1997; McRobbie 1997) and class (McRobbie 1997: 74), and that environmentally-sensitive consumption behaviours are both class and gender-sensitive (Harrison *et al.* 1996; Bedford 1999). However, for the remaining groups, our major concern was to ensure that a wide range of sectors were represented in the research. The aim here was to ensure maximum homogeneity within groups to enable us to gain a better understanding of the concerns of that sector.

Recruiting members for focus groups is not easy. For undergraduate dissertations, it is likely that a *snowballing* strategy would work best. Here, personal contacts are mobilized to bring friends along to the group (easiest if you are working with a student population or a group to which you already belong, such as a sports club, church group, etc.). Gatekeepers in the community or professional world can also be mobilized to pass the researcher along the network of contacts. Wherever people are gathered together can be helpful – one member of our research unit used the school run[1] to recruit people into a study of car use in Cambridge. In all cases, it is best to have a time and place arranged for the focus group, so that the individual can be sure about his or her availability. It is generally seen as standard research protocol to be entirely open with the recruits about the topic of the focus group (Kitzinger and Barbour 1999). However, care needs to be taken to ensure that the findings are not skewed by those who agree to attend the focus group having a certain interest in the topic. It is helpful to use a small questionnaire at the recruitment stage to ask questions about relevant opinions or demographic details.

Funded focus-group research will normally use a recruitment company to find participants who will be selected from random, short screening questionnaires in different locations – and this is how the consumer groups were recruited for the BRC study. Recruiting the professional groups for the BRC research was much more difficult. Writing to professional bodies and individual companies and following up with telephone calls were extremely time consuming, and, inevitably, the companies who agreed to take part in the focus groups tended to be self-selective rather than representative. Unlike the situation for lay publics,[2] it was more successful to have a variety of potential dates for the professional groups, and then to choose the date that the majority of the group members could attend. Given the active support of the BRC and assistance from their research officers, the best rate of recruitment came from the retailer groups, who were recruited from an existing environmental forum.

Recruiting pre-established groups with an interest in the subject being researched can make recruitment easier (Holbrook and Jackson 1996). The group is likely already to have a certain dynamic and set of norms that determine the pattern of the discussion (Kitzinger and Barbour 1999). This allows the researcher to 'eavesdrop' on the sort of conversations a group would be likely to have in a non-research setting (Holbrook and Jackson 1996). However, the group dynamic may well leave the researcher struggling to understand in-jokes and tacit group knowledge (Krueger 1994). We had mixed feelings about using pre-existing professional groups for the BRC research. Because the retailer representatives were all recruited by the BRC from their own environmental forum, this meant that many of the group members knew each other and the working environment of each other's corporations. This inhibited anonymity and limited group members' ability to put a green spin on their company's practices. Whilst this gave us access to a large amount of information about the retailers' environmental responsibilities, it was an uncomfortable experience for some of the professionals. We, in turn, became uncomfortable about what was being said. One group, in particular, knew each other so well that we did indeed become 'eavesdroppers' on their conversation, as they discussed corporate issues openly, seemingly oblivious to the tape recorder and the facilitators alike. Only at the end of the group did the participants become aware of the level of detail that they had given to us. They asked for certain parts of the conversation to be removed from the record, which was done.

Whilst Kitzinger and Barbour (1999: 8) suggest that some researchers are happy to work with groups as small as three individuals, in our experience we have found that if the number of members drops below five the group loses some of its dynamic. Our preferred number of group members is around eight. Ensuring that the right number of group members turn up for the meeting is extremely difficult. Drop-out rates are a very important consideration in recruiting for focus groups – even when people have promised to come and, in the case of lay publics, when an incentive payment is made. In the BRC research, the groups of professionals had a very high rate of not turning up on the day – a sudden spate of 'urgent, unscheduled meetings' among those who told us on the day of the meeting that they would not be coming to the group; or, worse, people simply not appearing in the meeting room. This is doubly difficult, because it makes those people who do come feel more uncomfortable. Academics who have recruited using pre-existing social and/or interest groups suggest that this decreases the drop-out rate for lay publics (Holbrook and Jackson 1996). If you are going to recruit using random sample methods, however, we suggest that you recruit between two and four more members than you would ideally like. Telephoning a couple of days in advance to check that members are still coming is also helpful, but is not a complete safeguard against non-attendance.

## Group conduct

Focus groups can be run in a variety of venues, from the front room of a group member to a local hall or university committee room. The setting should be appropriate for the subject under discussion and the acoustics suitable for tape recording. For lay publics, we tend to seat the group in a circle without a table (see Fig. 8.1). However, over the course of the BRC research, we found that professional groups were more relaxed when seated around tables. We normally offer group members some form of refreshment on arrival to develop an informal atmosphere, and if there is a cash payment to be made we like to give it casually at this point to get the formalities out of the way. We always give group members a badge, generally with only their first name on it (although some older members may be uncomfortable with this, and it is best to ask what name they would prefer if it appears to be an issue). The badge provides a sense of anonymity, whilst allowing the facilitator and other members to address people by name – it is difficult to remember names of around eight members from the brief introduction.

The focus group is normally started by a short introduction of the research being carried out and reasons for the group discussion. This is followed by each member

**Fig. 8.1** *A focus group in action*

introducing him or herself, which not only allows a taped record linking voices to names, but begins the process of developing a group identity. It is useful, at this point, to give a brief list of the *ground rules* of the group. Rules include an agreement not to identify individuals with comments after the group has finished. This allows anonymity and therefore more freedom of discussion. Most importantly, the researcher must make it clear that no one's real name will be attached to specific quotes in analysis and writing through the focus group. Anonymity is essential and must be guaranteed. To enhance the group dynamics, members are asked not to make notes during the discussion. Finally, the group should be asked not to talk over other people and to allow everybody to express their opinions. It is imperative to introduce the rules in a chatty and relaxed manner. Nothing kills a group dynamic more effectively than making the members too scared to speak in case they break any of the rules.

It is good practice to begin the group with a positive topic that will allow every member to speak, and in an affirmative way – for example, asking consumers what they enjoy about shopping. Equally, in winding the group up at the end, try to have positive things to say so that people leave feeling good about having given up their time to help you in your research. Focus groups should always start and end on time – even if the members appear to be enjoying themselves as the discussion nears the end. In winding up, remember to ask the group if they have anything to add, so that members have the opportunity to say anything that they felt was important but have not had the chance to express. Finally, the group should be thanked, and the facilitator should once again reiterate the fact that the discussion will remain anonymous.

The groups are run using a schedule designed in advance. Schedules work through more detailed sequences of questions and/or specific tasks for members to do (see e.g. Macnaghten et al. 1995). The schedule determines how the topics for discussion should be introduced, and we tend to have a rough list of the time available for each part of the discussion. In practice the timing is very flexible, but having developed a relative weighting for the balance of the discussion, the facilitator can have an idea of what else needs to be discussed in depth before the end of the group. In the BRC research, the agenda was as follows:

- What do the different sectors (consumers, retailers, suppliers, etc.) feel they are doing to address environmental issues?

- What do they feel that other sectors in the commodity chain are doing to address environmental issues?

- Who are perceived as having the most to gain and lose from putting in place environmental initiatives and legislation?

- What interaction have they with the other sectors that allowed them to draw these conclusions?

Much of the literature on focus groups talks about the need for a trained and skilled facilitator to lead the group (see e.g. Krueger 1994). Academic writing is beginning to question how essential it is for a researcher to be fully trained in group skills to be able to conduct effective research (Goss 1996: 119; Kitzinger and Barbour 1999: 13). In our own experience, having worked with a substantial number of facilitators, including graduate students learning their craft, we have yet to find one who has run a group so badly that the findings had no relevance. Whilst facilitators obviously hone their skills with experience, the desired qualities are those possessed by the average undergraduate – the ability to listen, the ability to think on your feet, and a knowledge of and interest in the subject the group is discussing.[3]

Having already developed a group schedule, the facilitator is aware of what he or she would like the group to discuss. The conduct of the group is, therefore, mainly about deciding when to intervene to stop the group discussion from wandering too far from the subject area or to ask a follow-up question to pursue an interesting topic of conversation; when to intervene to prevent disputes or to quieten dominant characters; and when to reinvigorate or redirect flagging conversation. In general, this requires the facilitator to be constantly thinking of probes and cues fully to explore potentially important topics of conversation (Stewart and Shamdasani 1990: 95), or to think of ways of introducing a new theme without upsetting the momentum of the discussion. The most important thing is to find a balance between allowing a discussion to develop along unusual, but potentially relevant lines, or intervening to keep the discussion on track (Kitzinger and Barbour 1999: 13). Rather than forcing yourself into a particular facilitator 'persona', it is best to be yourself throughout the group, as long as you take account of the effects of your style of facilitation when you come to analyse the transcripts (Kitzinger and Barbour 1999: 14). For example, were the group talking about a topic through their own choice, or because it had been repeatedly suggested to them through a series of interventions?

We always run focus groups with a facilitator accompanied by a participant-observer. Whilst the facilitator leads the discussion, the participant-observer is primarily responsible for watching the dynamic of the group. If members of the group are unable to bring their points into the discussion, the observer can offer them the opportunity to enter the discussion through a hand signal or verbal request that they speak. At the same time, the participant-observer can help rejuvenate a discussion by joining in the conversation, or even move the discussion along if the facilitator appears to be struggling. This supportive relationship between participant-observer and facilitator reduces the pressure on the facilitator to perform, and goes some way to stop the facilitator from worrying that he or she might lose control of the group. This is helpful, as running focus groups can be extremely draining emotionally. From experience, however, it is clear that different people have different styles of running focus groups. If you choose to work with a participant-observer, it is helpful to discuss what role you would like him or her to play prior to entering the group, and perhaps to agree some signals beforehand so that you can make it obvious when you do and do not require help.

## Debriefing, transcribing, coding and analysing focus-group data

Immediately after the group members have left the room, we take a few minutes to debrief: to discuss and record our thoughts on the group dynamics, problems encountered, topics covered and anything else we think is relevant to the analysis of the group discussion. To keep the group setting fixed in the researcher's mind we often draw a seating plan, showing where each member of the group was sitting. It is all too easy to forget people's voices, so the next urgent task is to produce a *running order* of the discussion. This involves listening to the taped discussion again (as soon as possible), and jotting down the names of each speaker in turn, with just a couple of words to mark the start of their intervention. The running order can then be used to guide the full or partial transcription of the full group discussion. Transcription is time-consuming and can be burdensome – but it must be done, and as soon as possible. Box 8.2 shows a brief extract from a transcribed group discussion. Key points to note are the use of sequential line numbers – to help locate specific pieces of text for further work and for inclusion in the final report; the indentation of text – to allow easy identification of speakers; and the use of very wide margins on the right-hand side of the page – to allow the analyst to make notes during coding. The transcripts themselves should record every detail of the conversation verbatim, and be read for accuracy whilst listening to the tape once more, to ensure that it is a faithful record of the group discussion.

## Box 8.2 Coded extract of transcript, showing formats

| Line numbers | Speakers and transcribed text | Notes |
|---|---|---|
| 1 | ANDY. If I can start by introducing myself. My name is | *Do I believe this?* |
| 2 | Andy, and I've been a member of the Allotment | *Cultural politics of* |
| 3 | Revolutionary Front for 6 years [*chuckles*]. We | *allotment groups?* |
| 4 | grow organic vegetables and try to persuade people | *Organic food* |
| 5 | they taste better than the supermarket stuff. | *clearly important* |
| 6 | SARAH. I'm Sarah. I live round the corner from the allotments. | |
| 7 | I've come today because I am worried about | |
| 8 | this, um, genetical modified stuff … [*agreement*] | *Not sure of the* |
| 9 | MODERATOR. so other people are also concerned about GMs? | *scientific term?* |
| 10 | JOHN. Well, I've read some recent reports in the paper, and | *Media influence* |
| 11 | these so-called experts, erm, working for Monsanta | *here?* |
| 12 | is it? They claim it's safe. But that's just rubbish | *Public* |
| 13 | because what's to stop the bees going from one field | *understanding of* |
| 14 | to another? [*pause*] Oh, By the way, I'm John and … | *science issue here;* |
| | | *skepticism about* |
| | | *expertise;* |
| | | *lay ecological* |
| | | *knowledge* |

Focus-group transcripts require some systematic form of analysis to capture the depth of material, and to avoid the manipulation of the findings to suit the researcher's own ends (Kitzinger and Barbour 1999: 2). We tend to analyse transcripts by line-to-line coding, using a set of theme codes, followed by the creation of a 'discursive' or 'mind' map of the discussion.[4] Fig. 8.2 shows an example of such a mind map from Burgess's study of perceptions of risk in woodlands (Burgess 1996). For the BRC research the codes were relatively straightforward and had been largely predetermined by the research questions. We coded for each sector's own environmental responsibilities, and the sources of their responsibilities; what they saw as other sectors' responsibilities; and the sources of their knowledge about other sectors' responsibilities. Finally we coded for where they thought responsibility would be best placed in the chain of provision and consumption. The mind maps ensure that the entire discussion is represented, as well as the specific coded quotes. This allows quotes to be kept within the context of the discussion, thereby limiting the misrepresentation of group material.

From the analysis of the transcripts and the set of debriefing notes, we wrote a full report on the research for the BRC, which concluded that each sector in the retail commodity chain was undertaking a variety of environmental responsibilities, although most of them found this a burden. All sectors thought that responsibility had been passed to them, when it was better placed with another sector, thereby revealing the

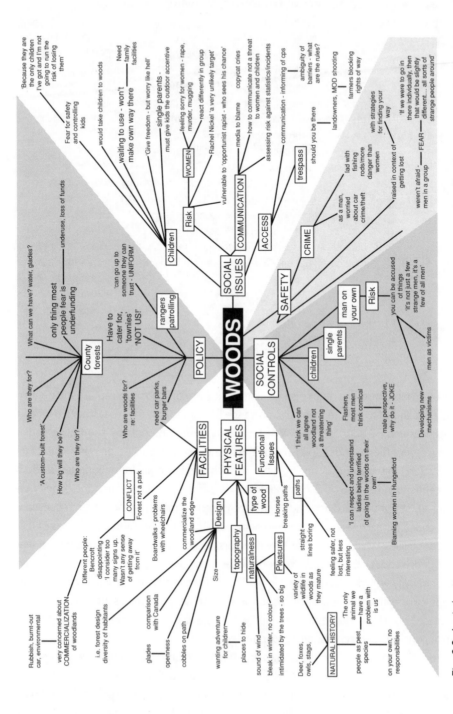

**Fig. 8.2** *A mind map*

complex interplay between the groups, as environmental responsibility is accepted and avoided throughout the chain of provision and consumption. The research showed clearly that the rhetoric of sustainable development had penetrated the commodity chain and had been accepted, but that the process of changing institutional and everyday practices remains a considerable challenge for both individuals and organizations.

## Conclusion

Hopefully this chapter will have persuaded you to think seriously about using the focus-group methodology in your dissertation research. In particular, it is worth considering using focus groups if your project involves collecting a variety of different groups' attitudes towards certain geographical locations, or their opinions about environmental or social issues. The technique can be used in conjunction with other methodologies, and is especially handy for establishing the range of issues that should be covered in interviews or questionnaires. It is a quick and effective method of gathering qualitative data, which has the advantage of uncovering lots of unexpected details about life that the researcher might never have previously considered.

In presenting the key phases of the methodology, and pointing out some of the issues that need to be addressed to ensure that research findings are robust, we have been eager to demonstrate some of the difficulties attached to running groups, as well as the benefits. To end, however, it is worth saying that running focus groups is great fun. Meeting different people, sharing in their often quirky and entertaining ways of understanding the world, listening to their stories and experiences – these are opportunities that no human geographer with any interest in how the social world works should pass by.

## Notes

1. Recruitment agencies are very effective at recruitment, but the high costs mean that their services are likely to be out of the realm of the average student.
2. By lay publics we mean groups of individuals who have 'everyday' knowledge of the subject under discussion, rather than expert/scientific knowledge.
3. Morgan (1998: 48) suggests that having a knowledge of the subject, and a long-term involvement in the project, is often more important for good focus-group facilitation than being an experienced facilitator.
4. For a more detailed explanation of qualitative analysis, see Strauss (1987) and Crang (1997), or see Cook and Crang (1995) for examples of coding. Morgan (1998) provides an overview of focus-group analysis strategies, with short examples from a variety of focus-group practitioners.

## Key references

Area 1996: 28(2).
Barbour, R. and Kitzinger, J. (eds) 1999: Developing focus group research: politics, theory and practice. London: Sage.

Morgan, D. 1998: *The focus group guide book: focus group kit 1*. Thousand Oaks, CA: Sage.

—— and Krueger, R. (eds) 1998: *Focus Group Kit*. London: Sage.

## References

Barbour, R. and Kitzinger, J. (eds) 1999: *Developing focus group research: politics, theory and practice*. London: Sage.

Bedford, T. 1999: Ethical consumerism: everyday negotiations in the construction of an ethical self. Unpublished thesis, UCL, London.

—— and Burgess, J. 1999: *Environmental responsibility in the chain of production and consumption*. London: British Retail Consortium.

Blake, J. 1999: Overcoming the 'value-action gap' in environmental policy: tensions between national policy and local experience. *Local Environment* 4(3), 257–78.

Burgess, J. 1996: Focusing on fear: the use of focus groups in a project for the Community Forest Unit, Countryside Commission. *Area* 28(2), 130–5.

Campbell, C. 1997: Shopping, pleasure and the sex wars. In Falk, P. and Campbell, C. (eds) *The Shopping Experience*. London: Sage, 166–77.

Cook, I. and Crang, M. 1995: *Doing ethnographies*. Concepts and Techniques in Modern Geography 58. Norwich: Environmental Publications.

Crang, M. 1997: Analysing qualitative materials. In Flowerdew, R. and Martin, D. (eds), *Methods in human geography: a guide for students doing a research project*. London: Longman, 183–96.

De Vault, M. 1991: *Feeding the family*. Chicago: University of Chicago Press.

Dowling, R. 1993: Femininity, place and commodities: a retail case study. *Antipode* 25(4), 295–319.

Goss, J. 1996: Focus groups as alternative research practice: experience with transmigrants in Indonesia. *Area* 28(2), 115–23.

Harrison, C. M., Burgess, J. and Filius, P. 1996: Rationalising environmental responsibilities: a comparison of lay public in the UK and Netherlands. *Global Environmental Change* 6, 215–34.

Holbrook, B. and Jackson, P. 1996: Shopping around: focus group research in north London. *Area* 28(2), 136–42.

Lunt, P. and Livingstone, S. 1996: Rethinking the focus group in media and communications research. *Journal of Communication* 46(2), 79–98.

Lury, C. 1996: *Consumer culture*. Cambridge: Polity Press.

Kitzinger, J. and Barbour, R. 1999: Introduction: the challenge and promise of focus groups. In Barbour, R. and Kitzinger, J. (eds), *Developing focus group research*. London: Sage, 1–21.

Krueger, R. 1994: *Focus groups: a practical guide for applied research*. 2nd edn. Thousand Oaks, CA: Sage.

—— 1998: *Analysing and reporting focus group results: focus group kit 6.* Thousand Oaks, CA: Sage.

Macnaghten, P., Grove-White, R., Jacobs, M. and Wynne, B. 1995: *Public perceptions and sustainability in Lancashire: indicators, institutions and participation.* Lancashire County Council.

McRobbie, A. 1997: Bridging the gap: feminism, fashion and consumption. *Feminist Review* 55, 73–89.

Morgan, D. 1998: *The Focus Group Guide Book: Focus Group Kit 1.* Thousand Oaks, CA: Sage.

Pringle, R. 1987: *Women, social welfare and the state in Australia.* Sydney: Allen and Unwin.

Stewart, D. and Shamdasani, P. 1990: *Focus groups: theory and practice.* Newbury Park, CA: Sage.

Strauss, A. 1987: *Qualitative analysis for social scientists.* New York: Cambridge University Press.

# 9
# Working with groups
## James Kneale

## Introduction

The group interview has become a well-known qualitative method as a consequence of the widespread use of focus groups in market research and policy formation (Kitzinger and Barbour 1999). Academic researchers have also found them useful (e.g. Lunt and Livingstone 1996) and an entire special issue of *Area* (1996) on group interviews is testament to a burgeoning interest in this method amongst geographers. Several useful 'how-to' guides exist for academics (Morgan 1988; Krueger 1994), for market researchers (Greenbaum 1998), and for both social-science and business practitioners (Morgan and Krueger 1997).

However, there are different kinds of group interview. Within geography the use of groups was pioneered by Burgess, Harrison and Limb at University College London in research conducted during the mid-1980s. This work rejected the market-research group tradition, adopting the principles and practices of group-analytic psychotherapy to explore the environmental discourses of lay people (Burgess *et al.* 1988a, b; Burgess *et al.* 1990). My own interest (Kneale 1995) in in-depth groups began with research into the imagined geographies produced by readers of science-fiction novels (see Fig. 9.1).

## In-depth groups and focus groups

In this chapter I shall be differentiating between *focus groups* of a more-or-less traditional sort and *in-depth groups* of the type conducted by Burgess and her colleagues. In-depth groups are distinguished by their relative longevity, with the group meeting several times;[1] their informal and supportive atmosphere; the limited intervention by the group's 'conductor'[2] and a consequent shifting of the power relationship between researcher and researched. Contributors to a recent collection have made similar claims for focus groups (Barbour and Kitzinger 1999), although their groups bear little resemblance to the market-research variant. In the following discussion I am comparing the in-depth group with the standard focus-group method.

In-depth groups provide an attractive methodology for three main reasons:

- the method recognizes the social nature of discourse;
- in-depth groups shift the balance of power away from the researcher;
- the working of the method allows for supportive meetings.

**Fig. 9.1** *Consuming SF: playing the Star Trek card game (photo: Kate Leyshon)*

### The social nature of discourse

In-depth groups produce dialogue within group relationships. Experiences and opinions are shared and challenged just as they are in other conversational settings. In justifying their positions, members examine and may change their own common-sense beliefs. Where groups meet more than once, these shifting agreements and disagreements create a 'group history'. A carefully transcribed discussion allows the conductor to situate each utterance in its context – is a statement a straightforward rejoinder, part of a long-running argument or support for a friend? I was particularly interested in group work for this reason, because I could see parallels with one of the theoretical interests of Mikhail Bakhtin's 'dialogical method' (see Holquist 1990). The dialogical nature of group work is well illustrated by De Maré's comment about group discussions: 'Every communication establishes attitude, role and relationship between individuals' (1972, cited in Burgess *et al.* 1988a: 312).

### A shift in the balance of power

Some of the researcher's control over the project is ceded to the researched. Group members have more freedom to choose the direction of the discussion within the constraints of the group relationship. In extreme situations, the group can literally shout the conductor down. Because the interviewees make their own choices as to what is relevant and interesting, group discussions can also unearth material that the conductor had not considered.[3] In addition, the interview transcripts are clearly polyphonic, though their translation into finished reports may reinstate the authority of the researcher (P.

Crang 1992; Duncan and Ley 1993). Because the conductor is also a member of the group, the in-depth group method encourages reflexive research practice: the researcher's preconceptions may be challenged and his or her comments are also part of the group dialogue.

In my interviews the groups generated material that I had not anticipated. However, several times I actually felt disempowered, because I could not respond to statements that I found offensive. For example, one group's discussion of difference relied upon a racist common sense that I would have disputed in other social situations. I chose not to contest it, partly because I did not want to intervene at this point but also because a group consensus had formed that was difficult to challenge.[4]

### Supportive meetings

As a group identity is formed and the members relax, the discussion often becomes humorous and supportive. The experience of participating in group work is often enjoyable; it gives the members confidence and in this sense the method can be empowering.

Nevertheless, no method is perfect and not all groups work perfectly all of the time. There are particular problems associated with group work: members can compete for 'leadership', or a consensus may silence the views of less confident members. I will come back to this later, but it should be noted that many of the problems of group work can be overcome by combining it with other qualitative techniques.

Now that I have established something of the nature of group work, the next section deals with research practice, drawing on my own experience to examine the practicalities of working with groups. My aim is to make the research process transparent by highlighting the practical problems and advantages of the method. However, while my research practice is considered in some detail, this is not a 'how-to' guide (see key references). The final section of the chapter examines the issues of group identity and conflict through an example from my own research experience.

## Research practice

The aim of my research was to explore the imagined geographies of science-fiction (SF) novels as constructed by ordinary readers. Three groups (referred to as A, B and C) were recruited and interviewed in October–November 1992 and May–June 1993. The first group (six members) met three times, and after some consideration I decided that this was not long enough to cover all the points I wanted. As a result the second and third groups (five members each) met four times. The research followed five stages: recruitment; preliminary interviews; group interviews; transcribing; and analysis.

### Recruitment

I began by choosing between homogenous and heterogeneous groups. Homogenous groups, where members are selected by age, gender or other significant characteristic, are common in market research. It is argued that 'the more homogenous the group is, the better the participants will relate to each other and the higher the quality of the input they will generate' (Greenbaum 1998: 62). In-depth group interviews can be used to explore the importance of social identities, and this can be an argument for grouping respondents on the basis of commonalities.[5] However, this raises the question of definitions of sameness and difference; a group that is homogenous in terms of gender, for example, might be heterogeneous in terms of age, class or other form of difference. Without wanting to duck the problems this raises, the key question is deciding which aspects of identity are most important to the research.

I considered recruiting an all-woman group, for example, because women are greatly under-represented amongst 'literary' SF fans, compared to media fandom, where they outnumber men roughly four to one (Jenkins 1992). This would have allowed these women a voice denied to them because of their exclusion from the masculine world of literary SF fandom. However, I decided that mixed-gender groups might allow a glimpse of how this exclusion is achieved. I was only able to recruit two women (compared to fourteen men), but fortunately their contributions did shed some light on the gendering of fandom.

The other common way of deciding membership is through interviewing already-existing groups, often 'friendship groups', which already have strong and supportive group identities.[6] These are easier to recruit, but the conductor will be a stranger in the group and the meanings of statements may be lost as a consequence.

Having considered these issues I took the decision to recruit mixed groups because 'differences between participants are often illuminating' (Kitzinger and Barbour 1999: 8). Recruitment was therefore structured by a desire to contact different kinds of SF readers, and took the form of:

- personal contacts;
- 'snowballing';[7]
- letters in two national SF magazines asking for respondents;
- fliers that were placed in a large Central London SF bookshop.[8]

Members of Groups B and C were offered £5 an hour for their time, chiefly to cover travel costs; this obviously helped, though several interviewees told me that they would have participated for nothing. The groups were filled as people came forward, though Group A was put together entirely through personal contacts and snowballing, Group B mainly through the magazine adverts and Group C mainly through the fliers. Each group contained friends recruited through snowballing.

The role of practical issues and chance should also be noted. One woman could not attend because the interviews took place in Central London in the evening and she was unwilling to travel at that time. Group A was larger than anticipated because a discussant brought his partner along unexpectedly.

### Preliminary interviews

The members of Groups B and C were interviewed singly before the groups met for the first time. This followed some awkward moments in Group A when discussion of personal tastes in SF began to break up the group dialogue. I also wanted to 'vet' potential discussants, as they had been recruited through more anonymous channels than the members of Group A. If I had felt that they would not enjoy the discussions or would make the others feel awkward then I would have subsequently contacted them to inform them that they would not be needed. This issue is not always relevant; it depends very much upon the nature of the discussion, the composition of the group and your own judgement as to the personal qualities of the interviewees.

The brief meetings and phone calls I had conducted with Group A were therefore formalized into interviews with potential members of Groups B and C. In these meetings – held in their homes or workplaces, or in cafés and pubs – I described the project and the group method, and recorded details of their media consumption. This was to provide myself with some contextual material to enrich my understanding of the group discussions.

### Group interviews

Each interview lasted around 90 minutes. I conducted them alone, but would have preferred to have used an observer to take notes and record non-verbal information (such as body language). The interviews were recorded for transcription.

I found successful conducting depended on careful use of a 'guide'. I prepared a filecard for each themed session, listing the topics that I hoped to discuss (see Box 9.1). The guide is used to 'track' the conversation, using prompts to maintain a smooth progression of points while simultaneously making sure that everything has been covered (Morgan 1988: 57). This is more difficult than it sounds, as it relies on maintaining the right level of conductor involvement. In in-depth groups the guide should be used to further or 'steer' the discussion rather than as a list of questions to be read out. If topics are left out, they can be introduced later on, but intervening too obviously risks calling attention to the role of the conductor.

Interventions can, however, be useful, and I risked some more formal techniques, particularly paper-and-pen exercises. I found these to be useful ways of breaking the ice in the first meeting and restarting discussion when the conversation had naturally broken down. Discussants were asked to write down three or four authors, themes or ideas, which were then offered to the group for general consideration. This allowed members to form individual opinions that were then discussed within the group framework, creating shared and contested views. The disadvantage of this technique is that it breaks up the flow of conversation, and briefly restores the convenor to centre stage. However,

**Box 9.1 Example guide**

Second meeting: technology

1. Has there been a decline in the status of science?

2. Has the importance of 'super-advanced' technology diminished in recent SF/cyberpunk?
   [Do we take it for granted? Does it no longer signal 'progress'?]

3. *Paper and pen exercise*
   Define 'robot', 'android', 'cyborg'.
   Which would you rather be?
   [Why? Can you imagine this existence?]

4. Virtual reality/cyberspace: what does it mean to you?
   [What does it look like? How would it be experienced?]

the discussion that followed was usually so successful that my intervention was quickly forgotten.[9]

Using the guides and exercises weakens any claim that these dialogues evolved without any input on my part. However, sometimes interventions are necessary. Conversations do run their course, unproductive silences, awkward topics and exclusive relationships do occur, and shifting the conversation by starting a new topic or using a paper-and-pen exercise is often a more sensitive way of dealing with these issues than calling attention to them. In any case, I am fairly confident that many of the interviewees were happy with these interventions.

Where the conductor has the assistance of an observer, they usually compare notes after the interview. I had to make do with my own memory of all the details of the discussion. Extra (non-verbal) context such as body language, glances, seating positions, mood, background distractions and so on were noted for inclusion in a research diary. In addition to the noting of these various elements, it is useful to 'brainstorm' ideas as to the significance of interesting aspects of the interview.

### Transcribing

Transcribing is a necessary but 'tedious' and 'tortuous' element of this method (Dahlgren 1988: 290). Burgess *et al.* estimate that 'a ratio of ten or twelve hours transcription to one hour of discussion would be realistic' owing to the difficulties of accurately capturing the conversation when people are arguing, talking over one another, or speaking quietly (1988a: 320). However, because individual utterances must be placed in their group context, this effort must be made. While transcribing I made short notes as to the meaning of certain statements or exchanges in my research diary. The transcriptions were given line numbers for quick reference (part of a transcript is reproduced in Box 9.2).[10]

## Box 9.2 Transcript of discussion B2

| Line numbers | Speakers and transcribed text | Notes |
|---|---|---|
| 1302 | MARK. Just to er … sound back over | |
| 1303 | a previous conversation, now it was an | |
| 1304 | obvious thing when you got to advanced | |
| 1305 | technology people became equivalent to | |
| 1306 | gods, you know, putting people on that | |
| 1307 | level, on that pedestal – | |
| 1308 | | |
| 1309 | [*hubbub*] | |
| 1310 | | |
| 1311 | ANNA. It's ridiculous. | |
| 1312 | | |
| 1313 | MARK. Yes, the local chemist is | |
| 1314 | going to be doing the genetic | |
| 1315 | engineering, and now there is no | |
| 1316 | technology, I can't think anybody, any | |
| 1317 | author who could seriously propose the | |
| 1318 | idea that this technology is so | |
| 1319 | advanced that we are gods, I mean the | *Argument* |
| 1320 | whole idea of gods – | |
| 1321 | | |
| 1322 | JACK. Erm, sorry – | *Correction* |
| 1323 | | |
| 1324 | DAVID. [*patronizingly*] Sorry, sorry. | *Backing Jack* |
| 1325 | | |
| 1326 | JACK. – Walter John Williams did it, | *Knowledge* |
| 1327 | [*to David*] in *Aristoi*? | *Establish link with David* |
| 1328 | | |
| 1329 | DAVID. Right. | |
| 1330 | | |
| 1331 | JACK. Yeah. | |
| 1332 | | |
| 1333 | DAVID. Dan Simmons does it in *Hyperion*. | *Knowledge* |
| 1334 | | |
| 1335 | | |
| 1336 | JACK. The thing is, Walter John's | |
| 1337 | following in the footsteps of his hero | *Contextualizing* |
| 1338 | Zelazny, and I wish he wouldn't. | |
| 1339 | | |
| 1340 | JK. So what's this technology that's – | *Intervention* |
| 1341 | | |
| 1342 | | |
| 1343 | MARK. Yes – | *Takes chance to speak* |

| | | | |
|---|---|---|---|
| 1344 | | | |
| 1345 | JACK. | [*mock disbelief*] You don't read | *Aimed at Mark and myself* |
| 1346 | | Walter John Williams? | |
| 1347 | MARK. | People [*laughs*] – people who | *Self-deprecating* |
| 1348 | | I read do not do it [*laughs*]. | |

## Analysis

Analysis began with the preliminary interviews and was supplemented by notes taken after each discussion and during transcription; this material was read alongside the transcripts and was then coded in much the same way as any other qualitative data would be (see M. Crang 1997). I then constructed 'discursive maps' (Burgess 1996: 133) to establish links and tensions between chunks of talk (see Fig. 9.2). However, because care must be taken to consider utterances as part of the group exchange, analysis must take account of the flow of conversation, changing relationships between group members and so on.[11] As a result, tools like discursive maps can be double-edged; discursive maps take utterances out of the flow of the conversation by sorting information by theme or code rather than the chronological order in which they came up in discussion. Imposing some sort of 'sense' on qualitative material always involves a loss of context, and this problem is greater in group work.

I have already noted that many of the particular strengths and weaknesses of group work concern the role of consensus and conflict. One of the key criticisms of the method, in fact, is that discussions may reflect the dominance of one or more discussant

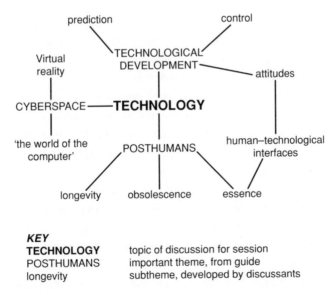

**KEY**
**TECHNOLOGY**   topic of discussion for session
POSTHUMANS   important theme, from guide
longevity   subtheme, developed by discussants

**Fig. 9.2** *A simplified discursive map*

or the polarization of debate in the form of an argument between two entrenched positions. I want to argue that these developments are not inevitable; that they can be useful research findings; and that these relationships are *precisely the point* of group work. I think it is useful to introduce these points with Kay Richardson and John Corner's comments on the relative merits of group work and one-to-one interviews.[12]

> *Our view is that, whilst both situations are clearly 'artificial' in a way that has to be remembered when using the talk as data, there are special difficulties with group work which suggest that research of this kind should involve a substantial element of one-to-one discussion, particularly in the early stages.* (1986: 487–8)

Richardson and Corner suggest that the main problem concerns 'the variables of domination, inhibition and consensus produced by group dynamics' (1986: 488), and this is a widespread criticism of group work. Krueger (1997) lists six types of 'people problems' (or problem people): 'experts', 'dominant talkers', 'disruptive participants', 'ramblers and wanderers', 'quiet and shy respondents' and 'inattentive participants'. Krueger's tips for dealing with these 'problems' are useful, but they might give the impression that groups are difficult to run or that they cannot be used to discuss sensitive issues (Morgan 1997).

These difficulties can be addressed in two ways. First, Richardson and Corner seem to suggest that group dynamics interfere with the process of group talk, while I would argue that no statement makes sense without reference to this dialogue. The question, surely, is not how to avoid consensus, but rather who dominates, who is inhibited and why does this happen? In my experience, only Group C worked in a generally consensual way, and the reasons for this were quite interesting. The group was dominated by three friends who shared a house and were studying the same college course; one was already clearly their 'leader'. All five members were very similar in age, background and taste, and, while there were disagreements, a particularly strong identity formed very early on. In other words, establishing that domination or consensus exists is a valuable research finding if we are at all interested in the dynamics of a group. It should also be remembered that these relationships are fluid ones, and may be negotiated and renegotiated in the context of ongoing group meetings.

My second response to these criticisms is that it is a question of research practice. In my groups I sometimes had to choose between intervening and allowing the group to find a consensus. Sometimes it was better to do the latter, and then watch what happened to this agreement; at other times I tried to make the participation in the group 'fairer', generally when someone wanted to contribute but was unable to. This is standard practice, and deciding which choice to make is an interview skill like any other.[13] It helps if you know the group well, and are able to evaluate the possible effects of intervention on the group relationship. The important point is that consensus is not necessarily something to be avoided, although it may be worth testing if it is becoming too solid. In fact, because some groups are valued for their consciousness-raising properties, the production of consensus (in the sense of shared understandings) might actually be an aim of the research.[14]

I want to explore some of these issues by looking at a moment when one of the groups did start to break down – though this is not an example of serious disruption, just the common-or-garden tensions that can emerge in any group conversation.

The five members of Group B (Mark, Martin, Jack, David and Anna[15]) were all in their thirties at the time of interview, except for Jack, who was in his early twenties. The older four all lived and worked in London; Jack was studying at a London college. All had science or IT backgrounds. David knew Anna from work and had recruited her. The group was dominated by the relationship between David and Jack, whose display of SF 'fan knowledge' marked them out as 'experts'. In the first meeting their shared interests and similar approach to reading ensured that they hit it off immediately. I had arranged the seating in anticipation of this, but without much success.[16] However, the first meeting was generally good-natured, as the discussants were getting to know each other (and me).

By the second session – which Martin was unable to attend – David and Jack's initial friendship had deteriorated into competition. To some extent this is to be expected at this stage of a group's formation. However, they were not contesting each other's leadership, but were trying to score points off each other by displaying their superior knowledge about SF. As a result a line was drawn between these two and Mark and Anna. At first I did not intervene because I wanted to see how the latter responded. When I decided that the situation was making them uncomfortable, I began to use eye contact and body language to move control over the conversation towards the other two. This worked to some extent with Jack, but David was oblivious. The following description of a discussion gives some impression of the nature of the relationships between the members at this point in the second meeting.

> We were talking about representations of technology in William Gibson's fiction. David and Jack were listing obscure authors and books, Mark continued to throw in comments, and Anna was clearly becoming peeved. Halfway through the session the tensions became obvious. In the following excerpt [see Box 9.2] Mark is arguing that the awe and respect given to science, technology and scientists in the SF of the post-war period is missing from contemporary science fiction.

Without any extra context, this looks like a description of a good-natured, if slightly pedantic, discussion. It is missing the following, which can be seen from the transcript in Box 9.2:

- the tone of patronizing correction in Jack and David's 'sorries';

- the redirecting of the conversation (the list of books and authors), so that it took a question from me to bring Mark back in;

- the way David and Jack use their more complete knowledge to draw a line between themselves and the rest of us (Jack's question towards the end of the excerpt was aimed at both Mark and myself), simultaneously authorizing their own opinions and establishing agreement between them;

- Mark's claim that 'people who I read do not do it [suggest technology makes us gods]' and his laughter suggesting he is aware that this has become just his opinion, based upon his narrower reading of the genre.

This exchange set the tone for the rest of the meeting. Anna and Mark began to disagree with David because of his attitude rather than because of what he was saying. I intervened several times by rephrasing questions or summarizing positions, not to take control of the conversation but to allow the others a point at which to break into David and Jack's dialogue. At one point Anna had to ask if it was all right for her to reply to one of Jack's comments and when David talked over her request I had to butt in. It was not until I introduced a paper-and-pen exercise that the conversation moved away from Jack and David.

At the beginning of the third session, I pointed out what was going on, and asked David and Jack to desist. There was an embarrassed silence until David acknowledged the issue and promised that they would stop. This intervention was rather awkwardly phrased but it seemed to do the trick. Anna was the first to speak; she spoke for a while, and when she handed the floor to Jack she did so with a joke about his youthfulness, a sly dig which positioned her as older and wiser than him while still maintaining the flow of the conversation. David waited for several minutes before making any comments at all.

To return to the points I was making at the beginning of this section, this situation certainly involved some discussants dominating the group and others feeling inhibited. If there was a consensus forming, it was one imposed upon the group by our two 'experts'. I didn't see it as a problem, however, until it became clear that Mark and Anna were irritated or upset by David and Jack's behaviour. In fact, I was fascinated, because one of the concerns of the research was the way in which readers talked about their reading practices.

David and Jack's game with each other – displaying knowledge and expertise – was interesting because they were the two members of Group B most closely bound up with the world of SF fandom. For these fans, this performance was a way of establishing their ability to attribute value to particular authors, books and ideas; their use of the terms of mainstream literary criticism also allowed them to borrow the legitimacy of the academy. Examining studies of fans of different kinds, John Fiske (1992) notes that some (mostly male) fans adopt this strategy to transform the weak currency of their fannish cultural capital into the real thing.

This consensus and conflict had therefore given me an insight into the ways in which this authority is performed in conversation, and the ways that it can exclude others who cannot display expertise or are unwilling to adopt the norms of approved cultural criticism to legitimize their opinions. I intervened because the playing-out of this conflict was ultimately damaging to the group, but it certainly was not a distraction from the real issues of the research – on the contrary, it provided me with some useful theoretical insights.

## Conclusions

I hope I have shown that in-depth group interviews are an extremely interesting way of working with complex issues. I also hope that my discussion dispels the idea that the conductor needs to be some kind of perfect 'people person'. There is nothing mysterious about conducting a group – there are plenty of tips that can be learnt, and we use these conversational skills every day without thinking about them. Most of my blunders were down to a lack of planning on my part.

My other main aim was to discuss the value of tensions in group discussions. Admittedly, in extreme cases disagreements can destroy a group. In most groups, though, episodes of conflict are inevitable and are as much a part of the research as the questions on the guide. This is not to suggest that the conductor should stir the group up and take notes while it degenerates into open argument. Discussants will look to the conductor for help if they feel they need it. In this sense, the group method, like all qualitative methods, requires balancing the needs of the conductor with those of the discussants: working with groups rather than setting them to work with you.

## Notes

1.  The number of meetings reflects the demands and practical limitations of the research, but the group relationship becomes richer with every meeting.
2.  I shall use the term 'conductor' to refer to in-depth group interviewers, and 'moderator' to refer to focus-group interviewers (Burgess et al. 1988a: 319).
3.  Agar and MacDonald (1995) stress this as one of the key benefits of the standard focus group.
4.  See Kitzinger (1994) for a similar case, and Wilkinson (1998, 1999).
5.  See Burgess (1996: 131–2) for an example.
6.  Holbrook and Jackson (1996: 137–9) discuss the problems of recruiting group members 'cold' and the advantages of finding established institutions and recruiters, which resulted in the recruitment of supportive friendship groups, the cooperation and backing of key figures and familiar places in which to hold the interviews.
7.  In each group one person I had contacted found one, two or three friends for me, so that each contained two or more friends.

8. I had hoped to find a key informant like Janice Radway's 'Dorothy Evans' (1984) in the owner of an SF bookshop in Central London and the editor of an influential SF fanzine, but neither could help.

9. Alternatively, I could have set the exercises for the group to do before the meeting (see Cook and Crang 1995: 62).

10. Transcriptions are identified by group, meeting number and line numbers from the transcript. The following notation has been used to represent the group discussions in print.

| | | |
|---|---|---|
| [*laughter*]   [*to David*] | non-verbal sounds or information, added for clarity |
| ... | short pause |
| — | break in discussant's statement |

11. See Myers and MacNaghten (1999) and Agar and MacDonald (1995) for conversational analysis of transcripts.

12. Richardson and Corner (1986) discuss focus rather than in-depth groups; their criticisms are applicable to both, even if they are much less problematic for the latter.

13. Morgan (1997) stresses the importance of the phrasing of questions; I found that prompts that dealt only in common-sense ways of thinking were unlikely to encourage the group to justify their statements.

14. Wilkinson's discussion of the value of group interviews for feminist research suggests that 'through realizing group commonalities in what had previously been considered individual and personal problems, women will develop a clearer sense of the social and political processes through which their experiences are constructed' (1999: 75).

15. All names are pseudonyms.

16. Talkative speakers should sit next to the conductor so that s/he can discourage them by leaning forward or turning away; quieter speakers should sit opposite the conductor where direct eye contact can draw them out.

## Key references

*Area* 1996: 28(2).

Barbour, R. and Kitzinger, S. (eds) 1999: *Developing focus group research: politics, theory and practice*. London: Sage.

Burgess J., Limb, M. and Harrison, C. M. 1988a: Exploring environmental values through the medium of small groups: 1. Theory and practice. *Environment and Planning A* 20, 309–26.

——  ——  —— 1988b: Exploring environmental values through the medium of small groups: 2. Illustrations of a group at work. *Environment and Planning A* 20, 457–76.

Krueger, R. 1994: *Focus groups: a practical guide for applied research*. 2nd edn. Thousand Oaks, CA: Sage.

Morgan, D. 1988: *Focus groups as qualitative research*. 2nd edn. 1997. London: Sage.

—— and Krueger, R. (eds) 1997: *Focus group kit*. London: Sage.

# References

Agar, M. and MacDonald, J. 1995: Focus groups and ethnography. *Human Organization* 54(1), 78–86.

Barbour, R. and Kitzinger, S. (eds) 1999: *Developing focus group research: politics, theory and practice.* London: Sage.

Burgess, J. 1996: Focusing on fear: the use of focus groups in a project for the Community Forest Unit, Countryside Commission. *Area* 28(2), 130–5.

—— Harrison, C. and Goldsmith, B. 1990: Pale shadows for policy: the role of qualitative research in environmental planning. In Burgess, R. (ed.), *Studies in qualitative methods*, II. London: JAI Press, 141–68.

—— Limb, M. and Harrison, C. M. 1988a: Exploring environmental values through the medium of small groups: 1. Theory and practice. *Environment and Planning A* 20, 309–26.

—— —— —— 1988b: Exploring environmental values through the medium of small groups: 2. Illustrations of a group at work. *Environment and Planning A* 20, 457–76.

Cook, I. and Crang, M. 1995: *Doing ethnographies.* Concepts and Techniques in Modern Geography 58. Norwich: Environmental Publications.

Crang, M. 1997: Analysing qualitative materials. In Flowerdew, R. and Martin, D. (eds), *Methods in human geography: a guide for students doing a research project.* London: Longman, 183–96.

Crang, P. 1992: Politics and polyphony: reconfigurations of geographical authority. *Environment and Planning D: Society and Space* 10(5), 527–49.

Dahlgren, P. 1988: What's the meaning of this? Viewers' plural sense-making of TV news. *Media, Culture and Society* 10, 285–301.

Duncan, J. and Ley, D. 1993: Introduction: representing the place of culture. In Duncan, J. and Ley, D. (eds), *Place/culture/representation.* London: Routledge, 1–21.

Fiske, J. 1992: The cultural economy of fandom. In Lewis, L. (ed.), *The adoring audience: fan culture and popular media.* London: Routledge, 30–49.

Greenbaum, T. 1998: *The handbook for focus group research.* 2nd edn. London: Sage. First published 1987.

Holbrook, B. and Jackson, P. 1996: Shopping around: focus group research in north London. *Area* 28(2), 136–42.

Holquist, M. 1990: *Dialogism: Bakhtin and his world.* London and New York: Routledge.

Jenkins, H. 1992: *Textual poachers: television fans and participatory culture.* London and New York: Routledge.

Kitzinger, J. 1994: The methodology of focus groups: the importance of interaction between research participants. *Sociology of Health and Illness* 16(1), 103–21.

—— and Barbour, R. 1999: Introduction: the challenge and promise of focus groups. In Barbour, R. and Kitzinger, S. (eds), *Developing focus group research: politics, theory and practice*. London: Sage, 1–20.

Kneale, J. 1995: Lost in space? Readers' constructions of science fiction worlds. Unpublished Ph.D. thesis, University of London.

Krueger, R. 1994: *Focus groups: a practical guide for applied research*. 2nd edn. Thousand Oaks, CA: Sage.

—— 1997: *Moderating focus groups: focus group kit 4*. London: Sage.

Longhurst, R. 1996: Refocusing groups: pregnant women's geographical experiences of Hamilton, New Zealand/Aotearoa. *Area* 28(2), 143–9.

Lunt, P. and Livingstone, S. 1996: Focus groups in communication and media research. *Journal of Communication* 42, 78–87.

Morgan, D. 1988: *Focus groups as qualitative research*. 2nd edn. 1997. London: Sage.

—— 1997: *The focus group guidebook: focus group kit 1*. London: Sage.

—— and Krueger, R. (eds) 1997: *Focus Group Kit*. London: Sage.

Myers, G. & MacNaghten, P. 1999: Can focus groups be analysed as talk? In Barbour, R. and Kitzinger, S. (eds), *Developing focus group research: politics, theory and practice*. London: Sage, 173–85.

Radway, J. 1984: *Reading the romance: women, patriarchy, and popular literature*. Chapel Hill, NC: University of North Carolina Press.

Richardson, K. and Corner, J. 1986: Reading reception: mediation and transparency in viewers' accounts of a TV programme. *Media, Culture and Society* 8, 485–508.

Wilkinson, S. 1998: Focus groups in feminist research: power, interaction, and the co-construction of meaning. *Women's Studies International Forum* 21(1), 111–25.

—— 1999: How useful are focus groups in feminist research? In Barbour, R. and Kitzinger, S. (eds), *Developing focus group research: politics, theory and practice*. London: Sage, 64–78.

# PART 4

## Participant observation and ethnography

# 10
# Fieldwork in the trenches: participant observation in a conflict area

## Lorraine Dowler

## Introduction

*How did you ever manage to interview the IRA?* This is a question that I am commonly asked regarding my research in Northern Ireland. My response seems disappointing to listeners, since I have no stories about clandestine meetings in alleyways or being blindfolded and taken to an interrogation centre. Instead, my reply reflects a rather commonplace moment in time and space: *I met them in a pub.*

Uneventful encounters with one's respondents, even notorious ones, are the earmark of participant observation. This method involves living and working within a community, in an effort to understand people's everyday lived experiences (Cook 1997: 127). There have been many cases of social scientists *getting into trouble* while trying to conduct research in Northern Ireland. These stories range from an individual who, while administering a questionnaire, was escorted out of West Belfast by three hooded IRA men, to the shooting of a Belfast anthropologist who was suspected of being a spy (Sluka 1995). These frightening turns of events were the result of the Irish Catholic community being distrustful of outsiders.

When I first selected Belfast as the site of my research I thought that my Irish Catholic background would help alleviate some of the community's distrust of strangers. Since I looked Irish, I naively assumed I would be accepted more easily as an insider. However, I was viewed as an American, and my red hair and freckles did little to alter my status as an outsider (Dowler, forthcoming). Participant observers can never fully shed their status as outsiders; however there are some steps that one can take to incorporate oneself more easily into a community.

At the time I conducted my research, Belfast was still a turbulent and violent study area; therefore I chose to concentrate my analysis within one community.[1] I determined that my actions might be considered suspicious, by both communities, if I was to live in one area (Irish Catholic West Belfast) and interview members of a rival neighbourhood (Protestant Loyalist West Belfast). I lived with a family in the Divis Estates, an Irish Nationalist area at the entrance to the Falls Road that is considered to be one of the most violent housing estates in the UK (Sluka 1995). I was placed with this family through the efforts of a local Catholic priest whom I had contacted a year before my arrival. I explained the nature of the project and asked him if he could locate a family who would be interested in taking in a boarder. This family was the first in a series of gatekeepers who would not only facilitate interviews but at the same time advise me how to act in a way that would lessen the suspicion of both the Irish Catholic community as well as the British security forces.

My first encounter with the IRA[2] happened a few days after my arrival in Belfast. I went to a West Belfast pub with the daughter of the family I lived with and some of her friends. As soon as we arrived the women eagerly chatted about all the local gossip. The women laughed and teased each other as they flirted with the men they were interested in dating. At one point their laughter changed into a nervous giggle as a group of men and women sat at the table next to us. In a whisper Rosin quietly leaned toward me and identified the group as the IRA.

As fate would have it, a conversation ensued between the two tables, which is typical of the convivial atmosphere of Irish pubs. My American accent was a giveaway that I was an outsider, but in this case served as a catalyst to a discussion of why I was in West Belfast. The end result of this accidental meeting was that I spent the next eight months interviewing, or, for a lack of a better term, hanging around with, this group and others to whom I was introduced.

I now realize that up to that moment I had been part of the production of an academic discourse that examines violence at a global scale rather than enquiring about the everyday experiences of both the victims and the agents of violence. I had always felt more comfortable conducting the 'in-the-trenches' type of geography rather than the proverbial armchair type; however, this experience opened my eyes to the consideration that, despite all my resistance, I had indeed been living in an ivory tower. Having read all the scholarly material written on the 'culture of terrorism', I had developed many preconceived notions of the constructions of a terrorist identity. My understanding of a 'terrorist' was abbreviated, abstracted from media images and academic writing on political violence. Ironically, I chose to enter this area as a participant observer – a methodology that uncovers the everyday experiences of people's lives – and yet I was uncomfortable observing this group in an everyday setting. Witnessing the group tell stories about their jobs, children and recent holidays seemed all too familiar, too commonplace and too natural. I remember being surprised that one man was wearing a New York Yankees baseball jersey. I thought to myself, 'What a far cry from a black balaclava.'[3] It would be the first of many encounters that in hindsight were far more significant than a researcher simply feeling empathy for his or her respondents – a celebrated advantage of participant observation (Chrisman 1976). Instead, my discomfort stemmed from the realization that violence was commonplace.

The fear of comprehending violence could be one reason why, although there is much academic writing concentrating on war, there is very little written on the everyday individual experiences of the violence that war generates. Nordstrom and Robben (1995: 1) comment on this phenomenon when they write:

> The term 'ethnic cleansing' made us remember other times and other wars and made us realize that the place may be different, and the suffering unique, but that everyday life under war is at any place and any time confusing and full of anguish. This realization is so obvious that it is almost banal, yet why is this perennial chaos of warfare and the incomprehensibility of violence for its victims so seldom addressed in scholarly writings? Why do we find so many intricate studies about war and so few about human suffering?

This chapter will explore the specifics of employing participant observation in a particular context, that of war. As part of this examination I will focus on five specific areas that should to be addressed when employing this method in a conflict setting. The first part of the analysis deals with my first impressions of the ethnographic site. In the second part, I explain the strengths and weaknesses of participant observation. The next part addresses the strategies I used to interview and observe a community that is distrustful of outsiders. In the fourth part I examine my everyday relationships with strangers. Finally I consider the post-ethnographic site. Often the identities of ethnographers who conduct research in conflict areas are designated as heroic or even worse conflated with the dangerous images associated with their informants (Swedenburg 1995). Therefore this type of methodology can bring danger to oneself in the field but also to the researcher within the halls of academia.

## First impressions

West Belfast is an area of the city that has been hard hit by the political violence that began in August 1969 (Sluka 1989). The area just west of the city centre is divided by a Peace Line, a fortified concrete and steel wall that divides the Shankill Road community – a predominately British and Protestant area – from the Falls Road area – an Irish Nationalist and Catholic enclave (see Fig. 10.1).

While in the area I informally interviewed both men and women, some of whom ideologically supported the Irish struggle and some of whom participated via paramilitary involvement. Many of these interviews took place in a Nationalist prisoners' club. This club was located in the heart of West Belfast, and, even though it was set just off from the Falls Road, a main Belfast artery, it was not noticeable from the street. In order to get to the club one must walk down a darkened alleyway and there in a large open lot behind some local shops stands a barricaded building. The first time I walked up to the club I was struck by its carceral image. The small windows covered by steel bars, the sequence of electronic gates leading to the front door in conjunction with surveillance cameras all gave this place a fortress-like appearance.

As I mentioned earlier in the chapter, I inadvertently became acquainted with a former

**Fig. 10.1** *Falls Road area, West Belfast, 1998*

prisoner that first week in the pub. This former IRA member became the gatekeeper of this community and he facilitated my interviews with past prisoners. Initially I could only gain entrance to the club accompanied by him; however, eventually my presence in the club became routine and I was able to come and go as I wanted (Dowler, forthcoming). The first time I approached the club my mind filled with Hollywood images of terrorist hideouts and smoke-filled command centres. Ironically the first group of people I encountered was a group of men and women who were singing the Don McLean hit 'American Pie'. This same group later went on to have a lively discussion about which university they hoped their children would be accepted to. My mental picture of a terrorist hideout was shattered and quickly replaced with the very commonplace notion of a PTA meeting.

This was not the first time that I experienced such a radical transition of place in West Belfast. In 1991 the apprehension I first felt when entering the prisoners' club paled in comparison to the dread I experienced when I first arrived at Divis Flats. Actually my anxiety about living in Divis started even before I arrived at the complex. Prior to going to Belfast I travelled for a week in the Republic of Ireland. During that trip, locals often asked me why I was going to Belfast. I was intentionally vague about my research agenda in order to avoid becoming embroiled in a heated political discussion. For the most part people assumed that I would be living at the university, and, while everyone I talked with generally had positive views of Belfast, I was issued a general warning to stay clear of the hotspots. Most specifically, I was told to stay away from Divis Flats, because it was a dangerous IRA stronghold. My fear was then compounded when I arrived at the Belfast train station and a taxi driver's response to my request to drive me to Divis was, 'Aye love, I can't take you there, I would get bloody shot in that place'. Although architecturally the complex resembled public housing elsewhere in the UK, this structure, covered in anti-British graffiti, was indeed a trope of this conflict. As I walked up to the complex I was taken aback by small children yelling at me 'the RA forever'. These children, playing around a car engulfed in flames,[4] immediately recognized me as an outsider. At this point two young men came up to me and asked, 'Are you the American the church sent?' Although I was an outsider, my introduction to this area by a Catholic priest expedited my presence not only to one of expectation, but also to one of acceptance. This was the first step in my quest to destabilize the boundary between insider and outsider.

The two men carried my bags and escorted me to the flat that I would be living in. I had to grapple my way up the dark internal staircase. It was very difficult to see where I was going, because all the lights had been smashed by local vandals. In actuality the lack of visibility was a blessing, since I was walking on top of a foot of garbage;[5] however, it did little to mask the smell of urine that permeated the hallway. By the time I knocked on the door of the flat I was convinced I would not last more than a night in this place. My mind was racing with ideas of other projects that I might be able to complete from the safety of the Queen's University archives; then, at that moment, a smiling older woman answered the door and invited me into her home. Despite the dampness due to the lack of heat and hot water, her home was lovely. The flat was clean, and beautifully wallpapered and furnished. Not only did I stay the night but I stayed for the next three months and have returned to live with this same family for all my stays in Belfast.[6]

This radical spatial transformation demonstrates how preconceived and stereotypical notions of a place can be dismantled by way of participant observation. To the outsider, the flats were the embodiment of the negative images of public housing; to the insider, however, this was a place where people went about their daily lives, raising their families and taking pride in their homes. This major change in experience prompts me to think of Hannah Arendt's description of participatory research as '*pearl fishing*: one dives in not knowing quite what one will come up with' (Arendt, cited in Elshtain 1987: p. xi). For this reason, in this particular context, I believe that the strengths of participant observation greatly outweighed its weaknesses.

## Strengths and weaknesses of participant observation

Participant observation, along with such other research strategies as in-depth interviewing and ethnomethodology, forms part of a general type of data collection and interpretation known as qualitative research. Such research strategies are usually contrasted to quantitative research strategies involving larger data sets and limited knowledge of individual subjects. If quantitative methods imply detachment and minimal contact with one's subjects, then participant observation implies attachment, involvement and intense contact with them. No research strategy offers unproblematic access to informants. All have their strengths and weaknesses. Researchers must decide which strategy to employ given the nature of the community that they wish to study.

Having said that, let me list what I see as some of the strengths and weaknesses of participant observation. Among the strengths is the ability to study behaviour in its natural setting. It has been argued that creating such natural conditions makes for more reliable responses by putting informants more at ease (Western 1992). This leads to a second strength of participant observation: greater depth of understanding. By putting people at ease and by spending sufficient time with a small group of people, one can get to know them in much greater depth than if one encounters them as a stranger and hurriedly administers a questionnaire.

There are, however, a number of problems associated with participant observation. One of the most important problems ironically stems from the strength of the method, its in-depth quality. Because of the time spent developing depth and detail, the sample size is very small. The question that arises, therefore, is to what extent can one generalize from the small group that one has studied in depth? There is, of course, an easy answer. It is best to beware of generalizing from such a sample, in the absence of other methods of verification, such as interviewing different groups or archival evidence. However, the researcher who employs this particular method must recognize that what one gains in depth, one gives up in breadth. Another commonly cited problem associated with participant observation is a loss of detachment as a result of being intimate with a group of people over an extended period of time (Chrisman 1976). Such a loss of detachment is inevitable in that one becomes involved, either positively or negatively, more often ambivalently, with the group. Most importantly, this should not be taken to mean that because of this one will knowingly distort the data that one is collecting. If one does so, then we are no longer speaking of loss of detachment, but of research fraud. The final problem is one of danger to the group one is studying, by the revelation of intimate details of their lives that could be used against them, and of danger to oneself. Given the nature of the community that I wished to study, I believed that the strengths of participant observation greatly outweighed its weaknesses.

## Beginning

I conducted informal interviews with my respondents in homes, shops, pubs, prisoners' clubs, on street corners and in taxi cabs. In most cases the interviews were conversational, so while I did jot down some observations, it was not until directly after a meeting that I wrote extensive notes. When working with a community it is important to obtain a range of backgrounds in the selection of respondents, especially if the community has undergone years of civil strife. In the beginning of this chapter I was critical of media and academic productions of a 'terrorist' identity. However, it is important to recognize that the terrorist can also be constructed by insiders as well as by outsiders. For example, at the beginning of this chapter I discussed my first encounter with a group of people who were described to me as the IRA. They were identified by a group who have never been involved with any paramilitary activities. Nonetheless, after spending a short time with these individuals I learned that only half of them actually had been members of the IRA, and, of those, none was currently active. This distinction was critical to my examination of the vying discourses of Irish Nationalism and would have been lost if I had not attempted to interview people with a range of backgrounds. Interestingly, the family I lived with would identify themselves as staunch Irish Nationalists. In the evenings they would often sing Irish Rebel songs and tell stories about the British Security Forces storming the flats. They spoke warmly of the men in the IRA, who had protected them by patrolling their streets. However, when I started also to spend time in prisoners' clubs I discovered that some former prisoners did not feel like the iconic heroes that were immortalized in resistance songs (Dowler, forthcoming). Surprisingly, they felt alienated from the larger Catholic community. More to the point, when the family I lived with discovered that I was working with former prisoners, they became angry. They referred to the group I was working with *as men who had done time*, not worth associating with. This distinction was critical to my understanding of the everyday events of Irish Nationalism. To the outsider, the landscape was laden with iconic images of these men proclaiming a community in solidarity; however, by way of participant observation, I was able to unpack some of the vying discourses within Irish Nationalism.

As I mentioned earlier a critical problem in this type of research is one both of bringing danger to the group one is studying by revealing intimate details of their lives that could be used against them, and also of bringing danger upon oneself. This was especially true in a violent atmosphere such as that of the public housing estates in Belfast. For this reason, when writing up my field notes I changed all the names of my respondents and purposely glazed over the details about locations. I would write up the day's notes late at night and often into the early morning. Every morning I would print out two copies of the notes, delete the notes from the computer[7] and go directly to the post office to mail a copy back the USA. I did this upon the recommendation of a contact at the American Consulate, who was often asked to intervene when the security forces confiscated notes from American researchers. Understandably, it would have been devastating to my research if I were to have had my notes confiscated. More importantly, however, I was concerned that, even though I had changed the names of my respondents, the notes could still jeopardize their safety.

In an effort to avoid bringing danger to myself I made a decision very early in the study not to cross the Peace Line to interview anyone from the Shankill (Protestant) community. I was concerned that I would lose the trust of the Irish Catholic community if I was seen consistently crossing the line between these two enclaves. Additionally I felt that it was safer for me personally to stay in one area, as in the past researchers have been accused of being spies, which in the best case cost them their project and in the worst cost them their lives (see Sluka 1989).

As you can see, given the nuances of working with communities who are wary of outsiders, much of your time is devoted to trying to establish relationships based on trust. As previously mentioned, this is viewed as a weakness of participant observation, since your sample size, in comparison to other methods, will be small. However, this method's weakness is also its inherent strength. The depth of understanding of the everyday experiences – that is, individuals' thoughts, hopes and fears while living under constant fear of attack – could not have ever been ascertained utilizing other methodologies.

## Everyday relationships with respondents

At this point I would like to focus on one of the advantages of participant observation: the degree of openness in social relationships with members of the studied group. As I mentioned earlier, this openness not only generates a better atmosphere for gathering data, but also increases the researcher's empathy with the people among whom he or she works. However, there are very real friendships that occur between the researcher and the respondent, which can not only improve the research but may also complicate it (Chrisman 1976).

A problem often ascribed to participant observation is a loss of detachment that comes from being intimate with a group of people over an extended period of time. However, terms such as empathy and loss of detachment do not sufficiently detail the level of friendship that can develop between a researcher and a respondent. Illustrative of this was when some of my female respondents asked me to help them write their own stories about their role in this conflict. They intended to invite women from the Shankill Road (Protestant area) to participate in the event. I immediately agreed and viewed this as an opportunity to establish an exchange of knowledge between the researcher and the respondent. However, many of the men, who were also my respondents, felt that this project was privileging a feminist agenda, thereby threatening an Irish Nationalist solidarity. This presented a rather dubious situation for me, for to become involved in the project risked my relationship with some of my male respondents, while not getting involved jeopardized my friendships with some of my female respondents. I made a decision to help when asked but not to be the developing force behind the project. I felt to do so would be inappropriate and thrust my research agenda onto their project. Sadly, because of the women's work schedules and family responsibilities, the project never came to fruition.

On another occasion, while I was interviewing a woman about her role in this conflict, she confided to me the problems that it created for her marriage. When she asked my opinion about acquiring a divorce, it became clear she was no longer talking to me as a researcher but was instead confiding in a friend. Although this discussion generally informed my analysis of gender roles in this conflict, I did not incorporate any of her direct testimony into the project. I made a decision that, although the discussion may have originated in the form of an interview, it concluded as a conversation between two friends.

There were countless other times when my friendships with some of my respondents risked impacting on the project in a negative way. Some examples include explaining to a respondent that it was not safe for her to be drinking alcohol while she was pregnant and stepping in front of a woman whose husband, also a respondent, was about to hit her. When conducting this type of research, there is no point of entry or exit to the interview. For this reason it is critical that the researcher make clear to his or her respondent what events and statements will be included in the manuscript. Each time I met a respondent I would consistently revisit testimony to ensure that he or she understood that our last conversation had been incorporated into my project. I adopted this policy after I had completed my first research project and sent back the thesis to obtain opinions from my respondents on the accuracy of the narratives. Although I feel I had been very clear about what statements would be included, they were still surprised and felt that some of the events I had written about were time simply spent 'hanging around'. However, despite this misunderstanding, I was still welcomed back for other research stays.

When a researcher becomes more friendly with a respondent, the researcher can usually behave in a manner that allows the respondent to feel more comfortable. However, no matter how well intentioned this action might be, the chances of the researcher misrepresenting themselves to the respondent increases. There has been a vast amount of academic writing focused on the issue of the proper representation of the researcher's subjects. Yet, as Katz (1996) points out, the way that researchers represent themselves to their respondents is of similar importance. Issues of personal politics and the researcher's willingness to be untruthful, for strategic reasons, are also of crucial concern.

While in West Belfast I represented myself in a way that would minimize my status as an outsider. For example, even though I am a staunch advocate of a woman's privilege to choose, I signed anti-abortion petitions. When I was asked how I felt about the issue of divorce, I was evasive. I promoted my Roman Catholic upbringing, even though I do not consider myself a practising Catholic. I attended weekly services at the local Catholic church and made sure I was seen at them. In hindsight, these actions did make my respondents feel more comfortable with me. However, I think it is important to ask ourselves whether, if we adopt participant observation in order to promote an open relationship with our respondents, the respondents should not have the same benefit of a candid relationship with the researcher? During later trips to Belfast I was far more forthcoming with respondents when they asked me about my personal beliefs. In the final analysis I believe that, although my respondents may have disagreed with me on some points, they were still quite willing to work with me; in fact, I would argue that my candour generated a richer experience for both of us.

## Post-ethnographic site

The process of writing an ethnography of violence became far more challenging upon my return home. Since many of my respondents had at one time been agents of violence, their testimony was challenged by some of my colleagues as not valid, the assumption being that, if respondents are involved in political violence, then they might not be credible informants. For example, a question that I am commonly asked is: *How do you know that these men are not lying to you?* This question's assumption of the sex, of my respondents as men, because of their association with violence, juxtaposed with the fact that I am a female researcher, presents an academic irony to me. Why should my male respondents lie to me any more than my female respondents would have done? Nevertheless, some academics seem to find my interviews with women far more credible. Some of my fellow male ethnographers who have worked within this same West Belfast community have applauded my interviews with women. They have expressed to me that it is fantastic that I am getting the 'unobtainable' woman's point of view; since I am a woman, I could 'naturally go where no male ethnographer had gone before'. The irony is that the women were far more reticent with me than the men ever were. In the club where I conducted the interviews, many of the women were suspicious of my intentions. I overheard one woman criticizing one of my male respondents for not checking if I was from the CIA or British Intelligence. I think in this particular context my gender disarmed the men from their usual suspicions of outsiders, while the women perceived me as a potential threat. Interestingly, some of my female colleagues did not focus on the 'natural feminine bond' between researcher and respondent that my male colleagues did; rather, they assumed that my exposure to agents of violence had in some way made me more 'manly'. For instance, shortly after returning from Belfast I was asked to give a presentation about my work at a Peace Studies workshop. After this presentation a female academic came up to me and quietly advised me to wear pearl studded earrings, as opposed to my brown beaded ones. She suggested that they might be appropriate for another type of research presentation, but that, because of the type of material I was presenting, I needed to appear more subdued. Similarly, another female academic advised me to wear 'very' feminine clothing when I presented my work. She counselled me to wear a skirt instead of slacks, so that I did not appear overtly masculine or militaristic.

Both these statements could be easily dismissed as not even worth mentioning; however, both of these women, as well as the male colleagues who challenged the trustworthiness of my respondents, point to the same phenomenon: the fear of comprehending violence. When a researcher employs a method such as participant observation, which requires getting to know the agent of violence in an everyday setting, there is a sense that the researcher has lost all objectivity and has become caught up in the romance of the conflict. It is a fear that the terrorist has some 'contagious magic', and that the researcher is contaminated simply through association and has become part of the revolution rather than observing it (Swedenburg 1995).

## Conclusion
I do not want to leave the reader with the impression that there is no danger involved in conducting research in a conflict area or that the agents of violence are simply

misunderstood people. To do so would be misleading, if not foolhardy. Rather what I hope to demonstrate is that Western epistemological analysis of war has centred on broad sweeping generalizations, which insulate academics from research that is grounded in 'people' and the way they experience the enactment of violence. Ironically, the interviewing of individuals who had violently taken human lives did not faze me, but my feelings of friendship for them were indeed unexpected. After my experiences in West Belfast I still may not condone violent measures of resistance, but I certainly understand why a society turns to them. I have had to look into a mirror and ask myself what I would have done under similar circumstances. As Feldman (1995: 225) admits, 'the ethnographic witnessing of terror asks of us without our intent or consent, to imagine however briefly, the position of the agents of violence'.

## Notes

1. The names of all my respondents have been changed in order to protect their identities.
2. There are many divisions within the IRA. This particular group identified themselves as the Provisional IRA, known locally as the PROVOS.
3. A black mask, which commonly covers the face of the terrorist.
4. Hijacked cars are often dumped and burned around the flats.
5. I later discovered that the residents threw their garbage in the stairwell in protest at the lack of proper garbage removal.
6. Although I did live with the same family, the Flats were razed in favour of two-storey housing, which I helped them move into at the end of my first stay in 1991.
7. I recently learned from one of my colleagues, who is an expert in the area of computer science, that merely deleting my notes, rather than reformatting the hard drive, would not have deleted my notes from the hard drive.

## Key references

Adler, P. and Adler, P. 1991: Observational techniques. In Denzin, N. K. and Lincoln, Y. S. (eds), *Handbook of qualitative research*. Thousand Oaks, CA: Sage.

Behar, R. and Gordon, D. 1995: *Women writing culture*. Berkeley and Los Angeles: University of California Press.

Cook, I. 1997: Participant observation. In Flowerdew, R. and Martin, D. (eds), *Methods in human geography: a guide for students doing a research project*. London: Sage, 127–50.

Nordstrom, C. and Robben, A. (eds) 1995: *Fieldwork under fire: contemporary studies of violence and survival*. Berkeley and Los Angeles: University of California Press.

Reinharz, S. 1992: *Feminist methods in social science research*. Oxford and New York: Oxford University Press.

Rynkiewich, M. A. and Spradley, J. P. (eds) 1974: *Ethics and anthropology: dilemmas in fieldwork*. New York: Wiley & Sons.

# References

Arendt, H. 1970: *On violence.* London: Harcourt Brace & Co.

Chrisman, N. 1976: Secret societies and the ethics of urban fieldwork. In Rynkiewich, M. A. and Spradley, J. P. (eds), *Ethics and anthropology: dilemmas in fieldwork.* New York: Wiley & Sons, 135–47.

Cook, I. 1997: Participant observation. In Flowerdew, R. and Martin, D. (eds), *Methods in human geography: a guide for students doing a research project.* London: Sage, 127–50.

Dowler, L. 1998: 'And they think I'm just a nice old lady'. Women and war in Belfast, Northern Ireland. *Gender Place and Culture* 5(2), 159–76.

—— forthcoming: Till death do us part: masculinity, friendship and nationalism in Belfast, Northern Ireland. *Environment and Planning D: Society and Space.*

Elshtain, J. B. 1987: *Women and War.* New York: Basic Books.

Feldman, A. 1991: *Formations of violence: the narrative of the body and political terror in Northern Ireland.* Chicago: The University of Chicago Press.

—— 1995: Epilogue: ethnographic state of emergency. In Nordstrom, C. and Robben, A. (eds), *Fieldwork under fire: contemporary studies of violence and survival.* Berkeley and Los Angeles: University of California Press.

Katz, C. 1996: The expeditions of conjurers: ethnography, power, and pretense. In Wolf, D. (ed.), *Feminist dilemmas in fieldwork.* Boulder, CO: Westview Press, 170–84.

Nordstrom, C. and Robben, A. (eds) 1995: *Fieldwork under fire: contemporary studies of violence and survival.* Berkeley and Los Angeles: University of California Press.

O'Connor, F. 1993: *In search of a state: Catholics in Northern Ireland.* Belfast: Blackstaff Press.

Sluka, J. 1989: *Hearts and Minds, Water and Fish: Popular Support for the IRA and INLA in a Northern Irish Ghetto.* Greenwich, CT: JAI Press.

—— 1995: Dangerous anthropology in Belfast: reflections on managing danger in fieldwork. In Nordstrom, C. and Robben, A. (eds), *Fieldwork under fire: contemporary studies of violence and survival.* Berkeley and Los Angeles: University of California Press.

Swedenburg, T. 1995: With genet in the Palestinian field. In Nordstrom, C. and Robben, A. (eds), *Fieldwork under fire: contemporary studies of violence and survival.* Berkeley and Los Angeles: University of California Press.

Western, J. 1992: *A passage to England: Barbadian Londoners speak of home.* Minneapolis: University of Minnesota Press.

# 11
# Multiple methods and research relations with children in rural Bolivia
## Samantha Punch

*Doing research is a messy affair, as dependent on negotiation, adjustment, personal choices and serendipity as on careful and meticulous preparation.*

(James *et al.* 1998: 169)

## Introduction

This chapter is based on ethnographic research carried out in a rural community, Churquiales, in southern Bolivia (S. Punch 1998). The study focused on children's negotiation of their autonomy at home, at school, at work and at play (S. Punch 2000, 2001). During the fieldwork, I lived for two extended periods in Churquiales (consisting of regular short visits over two years and a six-month intensive period of fieldwork[1]). I used a range of qualitative methods, including informal and semi-structured interviews and semi-participant observation with most members of a sample of 18 households. Full participant observation with children is impossible for adults, mainly because of their physical size (Fine and Sandstrom 1988), and it has been suggested that a semi-participant observer role is more suitable (James *et al.* 1998). It is this role that I pursued in Bolivia, as it enables the researcher to participate in children's activities to a certain extent whilst recognizing that there are limits to such participation. For instance, I could join in their games and ask them to teach me how to play, but I was a 'different' player, who was given special attention by the children, since adults do not usually play with them.

In addition, I spent three months carrying out classroom observation and task-based methods at the community school with children aged between 6 and 14. The school-based research consisted of children writing diaries, taking photographs, drawing pictures, completing worksheets and creating spider diagrams and activity tables. These last two methods were adapted from Participatory Rural Appraisal techniques (see Slocum *et al.* 1995; *PLA Notes* 1996). For example, the aim of the activity tables was to discover the range of activities and work that children do. They filled in a list of all the agricultural, animal-related and domestic tasks that they knew how to do, indicating whether they enjoyed doing that particular activity or not, and whether the activity was seasonal or year-round.

This chapter focuses on the building and negotiation of research relations, particularly with children, and on the use of multiple methods, particularly integrating semi-participant observation and informal interviewing. The first part looks at the development of relationships in the field, and considers the difficulties and implications not only of being an adult researching children, but also of being an outsider studying a different culture. It also explores the reflexivity and negotiation of the researcher's role. The second part of this chapter discusses the advantages of combining methods for this kind of research with children and adults in a different cultural environment, including a discussion of how such data were recorded.

## Being an adult *and* an outsider: an ethnographic and reflexive approach

The study had to address the implications of doing research with children, not only as an adult, but also as an outsider. A particular difficulty of conducting research with children of a different culture to the researcher's own is that a Minority World[2] experience of childhood involves understanding childhood in a particular way. The model of childhood in the Minority World is as a time devoted to play and school, free from many of the responsibilities of adult life. In order to understand rural childhoods in a country of the Majority World, preconceptions concerning personal experience needed to be minimized, along with notions of how childhoods are or should be. Nevertheless, studying childhoods within a different culture can in some ways be easier than studying the childhoods of one's own culture. This is because assumptions about those particular childhoods may not be so strong and the distance between the two cultures may facilitate the reflexive process. As Fetterman suggests, 'sometimes a familiar setting is too familiar, and the researcher takes events for granted, leaving important data unnoticed and unrecorded' (1989: 46).

Ethnography was the most appropriate research strategy for such a study for several reasons. As a white, middle-class female brought up within an urban environment in the Minority World, my background differed significantly from those whom I was studying. Naturally my background has created biases that can never totally be abandoned and perhaps should not be: 'Despite best efforts to suspend judgement and disbelief, who one is, what one believes and does, implicitly and ineluctably shapes the process and products of research' (James 1993: 8). Despite the cultural, social and economic differences between myself and the participants of the research, by living in the community for an extended period of time, I could become closer to their lives and closer to an adequate understanding of their culture and lifestyle.

The ethnographic approach accepts that researchers may influence the research context, since they become part of the social world that they study (Hammersley and Atkinson 1995). Reflexivity is a vital part of ethnographic research, as participation in the social world being studied requires constant reflection on the social processes and the personal characteristics and values of the researcher, which inform the data generated as well as the subsequent interpretation and data analysis. It was important to maintain a record of observations throughout the research process, especially of the context and how the children reacted. The process of how the data are produced plays a vital part in interpretation (see Mason 1996). By analysing the ways in which the participants respond to the researcher's presence, possible distortions can be recognized and minimized though not totally eliminated (Hammersley and Atkinson 1995). A reflexive approach is especially important in research with children: as an adult researcher, I needed to confront my own attitudes towards children as well as my role as an adult in a research process with child subjects.

Ethnography is a practical approach to employ when, as in this case, the research setting is in a fairly remote, not easily accessible rural area where daily visits to carry out interviews or observation would not be possible. The relative geographical isolation of the rural community also meant that many of the participants, especially the children, had experienced limited contact with outsiders. The children particularly tended to be very timid and unaccustomed to social interaction with people from outside their community. Many of the children had never seen a white European before my arrival at their community. Consequently, at the start of the research they reacted with stares and nervous giggles. When I tried to talk to them, they would run away or hide behind their parents. Initial conversations involved single-word answers on their part until they became accustomed to my presence. The children only started to feel able to talk to me after I had visited their household several times, during which I began to form a relationship of trust with their parents. Living in the community and taking part in some of the participants' daily activities meant I could form a relationship of trust vital for gathering detailed data which were sensitive to their perceptions of their lives.

## Building relationships

The nature of the researcher's relationship with participants must be acknowledged throughout the research period, especially because researchers frequently worry about the appropriateness of their behaviour in the field, particularly their 'methodology, personality and morality' (Devereux 1992: 43). Over time, immersion into the local culture and the formation of closer relationships with research subjects can result in the researcher becoming so absorbed with their lifestyles and particular anecdotal details of their lives that it can become difficult to detach oneself and observe as an outsider. A conscious effort must be made to record emerging ideas, difficulties and changing relationships with participants. This section examines my field relations with research subjects, particularly with children, how they developed and changed over time, acknowledging some of the difficulties that had to be overcome.

I encountered a variety of practical problems in the field, including: having to adapt to the regional vocabulary; learning how to deal with vicious dogs; health problems; rising rivers in the rainy season and transport difficulties. Some of my personal attributes could be used to my advantage: that I was a young female alone meant that many older women wanted to look after me. Nobody seemed to feel too threatened by me, which they might have done if I had been an older, male researcher, since people might regard such a person as a government worker asking about land for tax purposes. However, an initial dilemma that had to be overcome was the rumour that foreign females stole Bolivian babies to take back and sell in their own countries. Once, when a 4-year-old girl was sitting on my lap, a neighbour warned the mother of the household to take her from me or one day I would steal her. Fortunately my relationship was sufficiently good with that household that they could laugh off her suggestions.

My constant questioning would occasionally frustrate a few of the participants, and I felt a need to 'give something back' as a token of appreciation for their time and patience (see also Francis 1992). For example, on my trips to the town I brought back requested items such as a bag of rice or coffee, or recharged batteries used for cassette recorders, or replaced used gas bottles. When I visited migrants in Argentina, I took letters to relatives and brought back their replies. Some unrealistic requests had to be refused, such as financial help to build defences in the river to protect nearby cultivated fields from flooding. Caution had to be taken never to promise to do something that I would not be able to fulfil, since many rural people have experienced disillusion with outsiders whose promises to improve their lives never materialized. From the beginning I emphasized that the project and my research were to be studies with no direct material benefits, although copies of the project findings, reports, maps, booklets and photographs would be left with the teachers and community leaders. Despite having formed relationships of trust, they were still pleasantly surprised when I did keep this promise, since it was the first time they had received written documents about their own community. As Dolores,[3] one respondent, said: 'many people have come here to ask questions but then they disappear and we never see them again, and we never find out why they asked us those questions.'[4]

Throughout most of the fieldwork, I lived with two households: the families of Marianela (10 years) and Dionicio (12 years). During my initial visits to the community I had been particularly welcomed and befriended by these families, and, when I returned to stay for a longer period, they both asked me to stay with them. I would stay for about two weeks with Dionicio's family and then one week with Marianela's, as this family lived nearly an hour's walk from the village square, which meant that other household visits were more difficult to carry out. Neither family would accept payment for rent, so I repaid their hospitality by bringing food or other gifts from my town visits. In both of these households, I slept in the same room as the family, preparing and eating meals with them, becoming more like a member of the household rather than a privileged guest, as I had been at the beginning of the fieldwork. This meant I could observe and participate to a great extent in their lives, often witnessing the children's activities from the moment they woke up until they went to bed. Staying with two different families was ideal, as it gave me access to observe their family life.

My relationship with both the mothers, Felicia and Dolores, in these two households became stronger as the fieldwork progressed. Both of these women were key informants[5] on whom I could rely to cross-check information given to me by others, to ask them about others, to ask them for advice and to ask about sensitive issues. They were a vital source of rich information, and they were also used to my persistent questioning. Our best conversations were in the kitchen while I helped them prepare food, or by the river as we washed clothes. Since these were female domains, it was where they felt most comfortable and in control.

Felicia's and Dolores's children also became important key informants as they gradually let me inside their private world. It took longer to form a good relationship of mutual trust with the children than with the adults, because of the unequal power relationship between an adult researcher and a child participant. I spent much time with these children: accompanying them on their daily tasks, playing with them, walking to school with them, observing what jobs they did (see Fig. 11.1), what songs they sang, what games they played and how they negotiated their relationships with their parents and siblings.

Finally they began to be more relaxed with me, and open up to me about their social world. Dionicio is a good example of how such relationships can change with time.

***Fig. 11.1*** *Close-up of boy with cattle. Observation of children's tasks, such as ploughing the land, increased my understanding of their skills*

## Dionicio

Dionicio is twelve years old and is Felicia's youngest son. At first he was very shy with me. He used to hide behind a tree or hover in the doorway but would not say anything in my presence. Gradually he felt able to be in the same room as me but his responses to my questions remained monosyllabic. A successful ice-breaker was the laughter over a game I had brought with me, 'Pass the Piggies', which I played in the evenings with Dionicio, his two brothers and sister. I then began to accompany them on their daily tasks, such as fetching water from the river and milking the goats. I found it difficult to keep up with Dionicio as he nimbly ran up the mountainside to round up the goats. He laughed at me as the 'amusingly inept adult' (Ennew and Morrow 1994: 64) because I did not know how to milk the goats.

When I began doing the school-based research I walked to and from school with him. At first he tried to embarrass me in class because he was more confident with me and knew me better than the rest of the pupils. He pushed and tested me in our new roles as teacher and pupil. Children often test adults (Fine and Sandstrom 1988) and: 'When children think you are a listener, they will initiate conversations with you. If they know you are slow to criticize or difficult to shock, they will talk to you about anything and everything.' (Houghton and McColgan 1995: 69)

I do not agree that children will necessarily talk about anything once a researcher has gained their trust, but gradually they will allow greater access to their secrets and their worlds as the social distance between adult researcher and child participant is lessened (Thorne 1993). In time Dionicio became more co-operative with me in the classroom and he was keen to openly demonstrate our relationship at school. He started buying me chewing gum or an ice lolly at breaktime and giving them to me in front of the other children. At first I felt awkward, as I was not used to children giving me things, especially when I knew that they had so little money, and I was unsure whether to return the gesture. Initially I did nothing, but then I decided that, if when he bought himself something at break he got two and gave me one, then I could do the same. Consequently, on the days when I bought myself something, I bought two and gave him one in the same open way as he did with me. The other children (whom I worried about at first) did not seem surprised by this as they knew I was living in his house and seemed to presume the link to be fairly natural.

I finally knew that our relationship had progressed sufficiently when he looked at me one day and said: 'Don't they have combs in your country?' I responded that they did and asked him why he wanted to know, and he replied: 'Well, why don't you use one then?' This confident, rather cheeky, remark amused me by its frankness (my fringe tended to be backcombed rather than brushed) and surprised me by its boldness. I was pleased that he felt confident enough with me to say what he thought. Once our trust relationship developed he began to give me a variety of information, especially on the way to and from school when we were alone. For example, he confided in me that several times he had taken some eggs which he had conveniently 'found' and sold them so he could buy himself some sweets. I often asked him about the other children at school, in much the same way that I used to ask his mother about other people in the community. Access to such information was only possible after having built up a relationship of trust over a period of time.

However, caution had to be taken over revealing the hidden aspects of children's lives to adults. When children told me things in confidence and asked me not to tell their parents, I had to be careful not to mention them to their parents or teachers. For example, once I saw two children going fishing and they told me that they were supposed to be looking after their mother's cows. They asked me not to say anything to their mother if I saw her. Consequently, when I saw their mother and she asked if I had seen the children, I chose to lie and say I had not.

Forming relationships of trust can take a long time and varies with different people. Some people only began to take me into their confidence when they realized that others had done so. Some people never lost their suspicions of me as the foreign visitor to their community and made it obvious that they would rather I did not visit them; I respected their preferences. Various signs indicated when I began to be more accepted by particular individuals. For example, some would switch to referring to me in the Spanish *tu* familiar form of 'you' rather than in the more formal *usted*.[6] Similarly, I knew our relationship was improving when I was invited inside the house for the first time or when I was invited to eat a meal with household members rather than separately at the 'guest' table.

Field relations were strengthened over time and, as I learnt more about the lives of both the children and adults in the sample of 18 households, they too learnt more about me. This shared knowledge enhanced a mutual, if unequal, relationship. In some respects it proved to be ethically uncomfortable for me because, as I gained their trust, they would open up to me more, enabling me to get richer data about their lives. Sometimes I felt I was manipulating our friendship in order to get good data (see also M. Punch 1986) and these feelings remain with me as communication with them diminishes over time and distance. Such close contact could also prove problematic as I found it increasingly hard to switch a very informal conversation to a more formal semi-structured interview situation to talk about particular issues, even though in most cases my unease seemed to be unnecessary. I wanted to carry out semi-structured interviews with the parents in order to cover the same kind of questions that I asked the children on their worksheets at school so that I could compare children's and parents' perceptions of rural childhoods. In order to compare their responses effectively, I had to write notes, which I felt sometimes disrupted the flow of a more informal interview, which would be recorded afterwards.

## Negotiation of the researcher's role

Before going to Bolivia, I hoped to do some classroom observation in the community school because it would be an opportunity to get to know many children and I could observe them in a child-centred environment away from their parents. However, the teachers' reaction to such a request was uncertain and I was worried that they might feel uncomfortable. The small size of classrooms, the little desks and chairs might mean that my physical presence would disrupt the classes, causing the children to stare and giggle. After several weeks I asked if I could sit in on a few lessons, and found that, despite my worries, the teachers were eager to have an extra pair of hands. I quickly assumed the role of teacher's assistant and found myself colouring in pictures, making decorations and sweeping the classroom floor. On my first morning, I was left in charge of the pupils while the teacher went home to finish preparing lunch and before long I was being left in charge of the pupils for whole mornings. When I suggested that I might prepare something for them to do, rather than rely on work set by the teacher, this was well received and I began devising research tasks (such as worksheets, drawings and diaries). This is an example of how ethnographic research relies upon opportunity and requires a high level of flexibility.

I began by asking the pupils to draw pictures of their lives in the community and taught them how to write a diary, recording what they had done on the previous day, from when they got up to when they went to bed. I did not expect the children to want to continue writing these diaries, as many of them wrote slowly and were not familiar with such a task. I was therefore surprised when all of the children in that class (22 of them) said that they would like to continue with the exercise. Over time some children wrote less and less, but over half continued their diary writing for more than two months. I collected the diaries each morning and sat at the back of the class reading them, which also proved to be an ideal observation opportunity, as I could observe the class at the same time. The diaries provided a wealth of information about the everyday, routine aspects of children's lives.

I also prepared worksheets for the pupils to complete on different aspects of their lives, some of which were closely related to issues explored in the other task-based methods. For example, questions on one worksheet about aspects of children's lives in the community complemented the drawings and photographs they had taken. Another worksheet was drawn up as a result of the activity table that children had completed. All the activities that they had mentioned were listed, and further columns were drawn up so that children could include who usually did that task in their household, who helped, who never did it, and at what age they learnt or could learn to do it. The worksheets allowed for more detailed information to be obtained on the issues that had been identified by the children as important in their lives.

From the town I brought paper, coloured pens, pencils, exercise books and a camera to enable them to undertake different task-based activities. I always carried a range of different materials so that I could replace teachers if they were absent, as I was regularly asked to do. The opportunities were many: one teacher was due to retire soon and welcomed a lighter teaching load, another was often absent as he lived some distance away and sometimes the roads were impassable and all the teachers took a day off each month to collect their wages from the town. When the teachers were present, I resumed my role as teacher's assistant and read the children's diaries whilst observing the class. I spent time playing or talking with the children during break and lunch time. When I joined in their games, both adults and children thought it was amusing, but some adults seemed to frown on such strange behaviour, as adults in Churquiales would almost never play with children.

The challenge was not to behave like a typical domineering 'adult', which was difficult at times. Children often tried to provoke me to see how I would react, but an attempt was made to understand them on their terms, and from their point of view, withholding judgement from an adult perspective. However, that was not always possible in a classroom situation, where to a certain extent some order had to be maintained as the teachers were relying on me. The disadvantage of adopting the role of assistant teacher was that this reinforced the power inequalities between myself and the children, which was precisely what I had been trying to minimalize (see also Morrow 1999). A balance had to be struck between being their friend, a teacher, a teacher's assistant, an adult and a researcher. I had to be flexible to switch between these different identities, but where possible I let the children decide what role they wanted me to play. For example, when they asked me to play with them at break time I would become their friend, but when during class they would ask me how to spell a certain word I would be more like a teacher.

However, not all children wanted me to play the same role in certain situations. The following extract shows how my relationship with 10-year-old Marianela not only changed over time but was negotiated and adapted to different situations.

### Marianela

The first time I met Marianela, she hid behind her mother when I arrived at their house, giggling in an embarrassed manner. She is the eldest daughter of one of the households where I stayed. With time, she became more relaxed as she taught me her games and asked me riddles. I accompanied her to feed the pigs, milk the goats, round up the cows, fetch water and wash clothes in the river. One evening when it was getting dark, I went with her to round up the cows. She suggested we climb a nearby hill to see where the cows were. At the top she immediately spotted them on their neighbour's land, but I could not see them until she pointed them out to me.

She took me up a steep slippery hill which had sharp drops that had to be jumped over. I was getting more and more nervous, thinking I could hardly see where I was going in the darkness, and I was afraid of falling down one of the holes. Marianela, on the other hand, was able to find her way in the dark without hesitation and fearlessly climbed the steep mountainside. 'Come on,' she said, 'You're not going to fall.'[7] I was not so convinced. She took charge; holding tightly onto my hand, telling me not to worry, that I would be all right. I felt she slipped neatly into an adult protective role while I was being the nervous child edging my way along tentatively. Holding her hand, and telling myself not to look down, I kept going despite knowing that if I did slip she would not be able to stop me and I would just have to let go. Somehow I felt better holding on to her hand. On reaching safety, I released her hand and realized I had been squeezing it quite tightly. She easily recognized which cows were theirs and which belonged to the neighbour's household. She went over to them, calling their names and throwing stones in their direction to make them move.

During this trip, she told me how she got on much better with her dad than with her mum. Her dad has told her to bring in the cows every day so she went without having to be reminded by her mum: 'That's why my dad gives me money, 1.50 Bolivianos (equivalent to about twenty pence) in the morning to spend at school, and whatever I buy I share with my brother and sister. He doesn't give them money because they don't help as much.'[18] Marianela explained that her mum got annoyed because her dad spoiled her, but not the other children. She took her anger out on Marianela, shouting at her, and sometimes hitting her. It was sometimes difficult for me to accept that the people I had grown to respect did not always behave in ways that I liked. Nevertheless, maintaining a certain professional attitude towards the research meant that I would have to try to remain impartial and maintain a certain distance. This, however, was not an easy process, especially when latterly I considered some of them as my friends.

Meanwhile, the relationship between Marianela and me had progressed substantially and gradually she began to confide in me. However, despite enjoying my company at home, asking me to join her with chores or learn her songs and games, at school our relationship was remarkably different. At school she kept her distance, treating me as she treated the teachers: greeting me with a polite, but rather distant 'Hello' in the morning. I respected her signals that she did not want her friends to witness our special relationship at school. Maybe her peers would have teased her for knowing the new 'teacher'. I let her define my role, and at school I responded to her formal 'Good morning' in the same tone which she showed me.

It was interesting that Marianela treated me in a way that was different from Dionicio when we were at school. In public she maintained her distance from me, whereas Dionicio was keen to demonstrate our friendship. These two different reactions made me realize how difficult it can be to know how to behave most appropriately with children. Being an adult, but wanting to do the right things in children's eyes, can be problematic. My solution was to follow the children's lead wherever possible, letting them decide how they preferred to negotiate our relationship in different contexts, and reacting to their behaviour towards me (such as with Marianela's aloofness at school and Dionicio's sweet giving). The most appropriate way for an adult researcher to behave with children is to try to understand the situation from their viewpoint, by listening to them, observing and reacting to their behaviour.

## Combining methods

Rural children in particular tend to be shy with outsiders and not used to conversing with other people (except at school or at home). Semi-participant observation provides a way to get to know children better and to build trust. It is also an ideal opportunity to carry out informal interviews and to talk about issues as they occur, turning conversations to certain topics of interest (see Fig. 11.2). In addition, by semi-participating with the children, I learnt by doing and reached a greater understanding of children's activities. For example, I could feel how heavy the water is that they have to carry, or how back-breaking it can be to harvest peas for a long time. I could also witness the special skills that children have acquired in order to be able to do certain tasks, such as how they can nimbly climb up cliffs, find their way in the dark, identify individual animals and round them up. Active participation increased the depth of understanding through doing as well as observing.

Since I was living in the community, I was able to accompany children on their daily chores and observe aspects of their daily lifestyle that they may take for granted, and might not consider mentioning in an interview. For example, if I had only asked children about the work they do, many activities would have been omitted, as they do not consider much of what they do to be 'work'. Observation was also useful for capturing the context of children's work and negotiations, as well as for allowing for 'the recording of multiple task performance' (Reynolds 1991: 76), when children carry out several activities at the same time. Also, there are most likely some differences between what

**Fig. 11.2** *Girls sitting by the stream. Many informal interviews took place by the river whilst washing clothes*

people say they do and what they actually do in practice, which is why it was necessary to include observation methods.

However, one of the disadvantages of semi-participant observation is that it can be carried out only with limited numbers of children, as it takes time to build rapport and a relationship of trust (Reynolds 1986). It is difficult to compare the different sorts of data obtained, since each situation is different and not easily comparable. It relies heavily on flexibility, making the most of opportunistic moments in the field, and it cannot be easily planned. Consequently, I felt it was important to use semi-participant observation to complement the other methods I used, including the interviews and the written and visual methods. For example, one benefit of using task-based research activities at the school was that many children could complete them simultaneously, so that I obtained information more quickly and for a greater number of children than I would have done using individual interviews or observation techniques (Boyden and Ennew 1997: 107).

During the fieldwork period I visited a sample of 18 households regularly. The aim of repeated household visits was to monitor the household livelihood strategy, and to carry out informal interviews about a variety of different topics, for which I had a checklist for discussion in my head. By regularly visiting the same sample households, not only is a relationship of trust built up, but accumulative interviewing (see Whatmore 1991) allows for a detailed history of the household to be formed. Repeat interviews over a long period of time also facilitate access to all household members, some of whom may be absent during initial visits. This permits a multi-perspective of the household lifestyle and contributes to a fuller picture of their present and past situation. Household visits were also used to carry out semi-structured interviews with 15 parents and eight grandparents about their own childhood and about their children's lives.

The main disadvantages of household visits were that they were time-consuming and imposed on participants' time and privacy. The visits lasted from half an hour to a whole day, depending on what the household members were doing and how well I was accepted into their home environment. Wherever possible, I tried to accompany household members on their tasks, in order to carry out informal interviews whilst minimizing disruption to their daily routine. For example, I would talk to the women in the kitchen while preparing food, or to various household members while helping to harvest or round up the livestock, or while helping to peel potatoes (see Fig. 11.3).

Most of the recording of data was carried out immediately after observation or informal interview to keep the interactions as unobtrusive as possible (see also Boyden and Ennew 1997: 149). During the semi-structured interviews notes were written to facilitate the more detailed writing-up afterwards, which included observations of the interview setting and the reaction of the interviewee to the questions and to the researcher. Wherever possible, verbatim quotations were recorded for use in the presentation of data so as not to lose the richness of the participants' own language and choice of words. However, during semi-participant observation it was difficult to record exact quotations, except in the two households where I was staying, as there were more

**Fig. 11.3** *Girl preparing vegetables. Participant observation with children intruded less upon their time, as we could chat while completing household chores*

opportunities to write notes. Since my first university degree was in Spanish and Latin American Studies, my fluency in the language enabled me to design the research tools in Spanish. Language competency was also essential for carrying out ethnographic research of this kind in the first place, since rapport and research relations could not have developed in the same way via an interpreter. Yet, despite fluency in Spanish, many local terms and farming vocabulary also had to be rapidly learnt. I translated all the quotations, but it was not always an easy task to capture the exact meaning in English. Consequently, my translation was sometimes flexible to incorporate the flavour of the local language, which was why I decided to keep the original words in the footnotes.

I chose not to use a tape recorder, mainly because it was not a practical option, since most of the informal interviews were carried out whilst accompanying the respondents during errands or tasks. During semi-structured interviews I did not want to make the respondents feel more self-conscious, especially since they are not used to being interviewed and are unfamiliar with tape recorders. Consequently, I developed my own form of shorthand, scribbled notes to prompt my memory and learnt to write quickly whilst maintaining eye contact and the flow of conversation (Boyden and Ennew 1997: 149).

Approximately every ten days I withdrew from the community to spend two days in the town of Tarija, to enable me to reflect on the data obtained and on my role as a researcher, to transfer field notes to a laptop computer and to consider how my ideas were developing. Copies of detailed letters sent home regularly to family and friends were kept, describing the nature of field relationships, the cultural differences, the uncertainties, joys and dilemmas of life in the field. These have proved useful in reconstructing the changing nature of the fieldwork and the intellectual process involved.

## Summary

This chapter has shown some of the advantages of combining a range of qualitative methods. Each method has particular advantages and disadvantages. To a certain extent, the questions and themes to be explored determined the methods chosen. For example, processes of negotiation between adults and children could not be depicted in a drawing, photograph or diagram, but needed to be observed, written or talked about. Practical issues also determined the choice of methods. Semi-participant observation and household visits were time-consuming and could be used only with small numbers of children, unlike the task-based methods, which could be carried out as classroom activities.

The visual methods of using drawings, photographs and PRA techniques were most useful in the initial exploratory stages of the research for the investigation of broad themes and for seeking children's definitions of the important aspects of their lives. The written methods of diaries and worksheets were used to examine those issues that children had raised in more detail. The household visits were useful to provide a broader perspective of their social worlds. They were used with semi-participant observation as complementary to the other methods, to confirm whether children did in practice what they said they did, and to add greater depth to an understanding of children's rural lives in different contexts.

In addition, this chapter has explored some of the difficulties of developing research relations in the field, the key ingredients in building such relationships being sensitivity and flexibility to adapt to the particular research context. The chapter has also focused on the negotiation of the researcher's role and the importance of switching between different identities according to the requirements of the research setting and the preferences of the individual research subjects. The researcher should be prepared to invest time and make the most of opportunistic moments to give something back to participants, allowing a more mutual relationship to develop based on trust. Ethnographic researchers should also be ready to carry out informal interviews as and when opportunities arise as well as being willing to fit in with participants' activities and daily lives. However, in order to maintain an awareness of the developing nature of research field relations, a detailed record should be kept not only of the researcher's changing thoughts, feelings and observations but also of the participants' reactions to the research process.

## Notes

1. This doctoral research developed as a result of initially working for two years in the same region of southern Bolivia on a European Union funded project titled: *Farmer Strategies and*

*Production Systems in Fragile Environments in Mountainous Areas of Latin America.* Both the EU project (managed by Dr David Preston) and the doctoral study (including a follow-up fieldwork period of six months) were based at the School of Geography at the University of Leeds. Grateful acknowledgement goes to the School of Geography at Leeds and the British Federation of Women Graduates for their financial support of the doctoral research.

**2.** I prefer to use the terms Majority World and Minority World to refer to the Developing and Developed World respectively. The Majority World has most of the world's population and a greater land mass. Thus, it reflects the majority experience compared to the more privileged lifestyles of the Minority World. Although this unduly homogenizes the 'Majority', by my using the terms Minority and Majority World at least the reader may be caused to pause and reflect on the unequal relations between these two world areas.

**3.** All the names of the respondents as well as the community have been changed in order to protect their identity and maintain confidentiality.

**4.** 'Mucha gente han venido aquí para hacernos preguntas, pero despues se disaparecen y no les vemos de nuevo. Asi no sabemos porque nos han hecho esas preguntas' (Dolores, Churquiales 1993).

**5.** 'Key informant' is used here as meaning respondents with whom close friendship was formed and much time was spent, and whose opinions were regularly sought (Boyden and Ennew 1997: 124–5).

**6.** In southern Bolivia, people tend to use *vos* instead of *tu* for the familiar form 'you.'

**7.** 'Vamos, no te vas a caer' (Marianela, 10 years, 18 Oct. 1996).

**8.** 'Por eso mi papá me da plata, Bs.1.50 en la mañana, y de lo que compro doy a mis hermanos. Pero él no da a ellos porque ellos no lo ayudan tanto' (Marianela, 10 years, 18 Oct. 1996).

## Key references

Devereux, S. and Hoddinott, J. (eds) 1992: *Fieldwork in developing countries.* London: Harvester Wheatsheaf.

Hart, R. 1997: *Children's participation: the theory and practice of involving young citizens in community development and environmental care.* London: Earthscan Publications Ltd.

Hill, M. 1997: Participatory research with children. Research review. *Child and Family Social Work* 2, 171–83.

Johnson, V., Ivan-Smith, E., Gordon, G., Pridmore, P. and Scott, P. (eds) 1998: *Stepping forward: children and young people's participation in the development process.* London: Intermediate Technology Publications.

Lewis, A. and Lindsay, G. (eds) 2000: *Researching children's perspectives.* Buckingham: Open University Press.

## References

Boyden, J. and Ennew, J. (eds) 1997: *Children in focus: a manual for experiential learning in participatory research with children.* Stockholm: Rädda Barnen.

Devereux, S. 1992: Observers are worried: learning the language and counting the people in

Northeast Ghana. In Devereux, S. and Hoddinott, J. (eds), *Fieldwork in developing countries*. London: Harvester Wheatsheaf, 43–56.

Ennew, J. and Morrow, V. 1994: Out of the mouths of babes. In Verhellen, E. and Spiesschaert, F. (eds), *Children's rights: monitoring issues*. Gent: Mys & Breesch, 61–84.

Fetterman, D. 1989: *Ethnography: step by step*. Applied Social Research Methods Series 17. London: Sage.

Fine, G. A. and Sandstrom, K. L. 1988: *Knowing children: participant observation with minors*. Qualitative Research Methods. Series 15. London: Sage.

Francis, E. 1992: Qualitative research: collecting life histories. In Devereux, S. and Hoddinott, J. (eds), *Fieldwork in developing countries*. London: Harvester Wheatsheaf, 86–101.

Hammersly, M. and Atkinson, P. 1995: *Ethnography: principles in practice*. London: Routledge.

Houghton, D. and McColgan, M. 1995: *Working with children: child care*. London: Collins Educational.

James, A. 1993: *Childhood identities: self and social relationships in the experience of the child*. Edinburgh: Edinburgh University Press.

—— Jenks, C. and Prout, A. 1998: *Theorizing childhood*. Cambridge: Polity Press.

Mason, J. 1996: *Qualitative researching*. London: Sage.

Morrow, V. 1999: 'It's cool, … 'cos you can't give us detentions and things, can you?!': reflections on research with children. In Milner, P. and Carolin, B. (eds), *Time to listen to children: personal and professional communication*. London: Routledge, 203–15.

*PLA Notes* 1996: *Children's participation* 25. London: International Institute for Environment and Development.

Punch, M. 1986: *The politics and ethics of fieldwork*. London: Sage.

Punch, S. 1998: Negotiating independence: children and young people growing up in rural Bolivia. Ph.D Thesis, University of Leeds.

—— 2000: Children's strategies for creating playspaces: negotiating independence in rural Bolivia. In Holloway, S. and Valentine, G. (eds), *Children's geographies: living, playing, learning and transforming everyday worlds*. London: Routledge, 48–62.

—— 2001: Negotiating autonomy: childhoods in rural Bolivia. In Mayall, B. and Alanen, L. (eds), *Conceptualising child–adult relations*. London: Falmer Press, 23–36.

Reynolds, P. 1986: Through the looking glass. Participant observation with children in Southern Africa. Paper presented at a Workshop on the Ethnography of Childhood, King's College, Cambridge.

—— 1991: *Dance civet cat: child labour in the Zambezi Valley*. Athens, OH: Ohio University Press.

Slocum, R., Wichart, L., Rocheleau, D. and Thomas-Slayter, B. (eds) 1995: *Power, process and participation: tools for change*. London: Intermediate Technology Publications.

Thorne, B. 1993: *Gender play: girls and boys in school*. Buckingham: Open University Press.

Whatmore, S. 1991: *Farming women*. Cambridge: Polity Press.

# 12
# Negotiating different ethnographic contexts and building geographical knowledges: empirical examples from mental-health research

Hester Parr

## Introduction

Human geographical research facilitates the investigation and analysis of both individual and collective social worlds in order to gain more in-depth understandings about society–space and people–place relationships. In common with much work in the wider social sciences, and in the light of recent postmodern and poststructuralist concerns, human geographical research has also been influenced by a deliberate shift to 'recovering' and 'centralizing' what might be considered to be 'marginalized' voices and stories through more intersubjective research practices. Marginality in human geography is defined widely, and geographers have been attentive to looking at poverty, class, ethnicity, disability, gender, sexuality and citizenship, as these axes of human difference are implicated in the production and understanding of marginality. Marginality can be understood as a state of human being that is partially 'outside' mainstream institutions, cultures, practices, beliefs and spaces. Researching what one might call 'marginal social worlds' has thus attracted much attention in human geography, as part of an attempt to understand how people who embody such categories as those mentioned above live out marginal(ized) everyday geographies. Such a focus has also led to new methodological considerations about how such complicated aspects of everyday life can possibly be researched, known and understood (England 1994; Gilbert 1994; Kobayashi 1994; Nast 1994; Staeheli and Lawson 1994; WGSG 1997).

In view of such discussions, it is perhaps pertinent to think through more carefully the strategies that are used to access and research the social worlds of so-called marginal others. Investigating and understanding 'marginal social worlds' is not an easy task, and a key point to make in the beginning is that understandings are always partial, and our geographical visionings will always be contingent upon the 'definitions of the situation' (Jackson 1988) as they emerge through particular research projects and locations. It is important to remember, however, that, whatever the definitions of marginality and marginal social worlds, research in this context involves investigating people who are in some ways vulnerable. This vulnerability may be related to their jobs, homes, social networks, health or state-of-being in the world. Whilst it is not advisable to make blanket assumptions about what does and does not constitute a 'vulnerable' or 'marginal' status in human geographical research, is it perhaps worth exploring some issues involved in the investigations of people who live out 'marginal social geographies', in order that we might recognize different possibilities, and also sensitive and (un)successful strategies that are sometimes used in this respect.

In thinking about such issues in this chapter, I am going to draw on research carried out in Nottingham during the mid-1990s concerning the social geographies of people with mental-health problems. People with mental-health problems can be considered as a marginal (although heterogeneous) social group. Historically, they have been deliberately located together within isolated special asylum spaces (Philo 1997). This has happened because society has sought deliberately to marginalize those whom it considered to be dangerously 'unreasonable', placing them away from mainstream and 'reasonable' populations (Foucault 1967). The geographical imagining that has informed the care and placing of such people in this way was reversed in the late twentieth century by the policy of deinstitutionalization in Western societies, where people with mental-health problems become supposedly cared for in mainstream community settings. However, as a group, people with mental-health problems are arguably still stigmatized by this past institutional legacy, and their everyday geographies are often characterized by poverty, isolation and limited socio-spatial spheres (Giggs 1973; C.J. Smith 1981; Kearns and Taylor 1989; Kearns 1990; Parr 1997a, b). Given this situating of such people as generally marginalized, it is appropriate that geographical research should be sensitive to these contexts, and that our methodologies reflect and take into account the ways in which mental patients' 'voices and stories' have been ignored, excluded and deemed 'unreasonable' in wider society. As has been argued in previous works (Parr 1998a, b), this means careful consideration of the 'psychodynamics' of intersubjective encounters such as interviewing, as well as thinking through the different possibilities for investigating this group ethnographically. It is this latter area that is the main focus of this chapter.

Ethnography refers to processes of participating and observing in particular spatial settings for the purposes of research, and provides a way 'to understand the world views and ways of life of actual people from the "inside", in the contexts of their everyday, lived experiences' (Cook 1998: 127). Referring to people with mental-health problems, this research method allows a closer relationship with people who might find it difficult to answer a questionnaire, or to concentrate for an hour or two in an 'in-depth' interview (partly because of a diagnosed 'condition', but also because of drug treatments (see Parr 1998a)). Also, more distanced surveying of this group can be argued to aid in a construction of such individuals as medicalized objects, unproblematically distributed through the city in predominantly 'sick locations' (Giggs 1973). The decision to research ethnographically allows for a less structured, more flexible frame within which to work, but this also means that research can seem 'messy', 'uncoordinated' and 'unproductive' of the 'right sort' of geographical information or knowledge. In order to discuss ethnographic methods further, the remainder of this chapter will unfold two quite personalized accounts of attempts to 'do ethnography'. These accounts are not meant to provide 'the right way' forward or a definitive 'how-to' guide, but rather they simply reflect some of the grounded processes and also very real difficulties of negotiating access in this sort of research, as well as providing some pointers as to how different sorts of ethnographic encounters might result in the production of different sorts of geographical knowledges. A key theme of the accounts presented here is that researchers' 'positioning' in ethnographic work is worth self-reflection, as this can help in the understanding of the geographies under investigation (following S. Smith 1988).

## Two ethnographic encounters: reflecting on the pathways to constructing geographical knowledges

### 'Blurry' ethnography: overt/covert research with a psychiatric service user collective

When my research project first began, I made contact, via letter, with an umbrella psychiatric service user group in Nottingham. This organization's main function is to support the formation of local 'user groups' in medical environments across the city, in order to facilitate user empowerment and user-led change for people with mental-health problems in places of psychiatric treatment (for a more detailed account of this group and its functions, see Parr 1997b).¹ Access to this organization, the 'Nottingham Patient Council Support Group' (NPCSG), was negotiated via a nerve-racking encounter with the group's development worker, who questioned me about what I wanted to know about the people in the organization. At first, I had no plans for this contact to become an ethnographic research exercise: I had wanted to contact the group in order to gain access to ex- and present patients with respect to carrying out in-depth interviews about themes such as isolation, stigma and marginality in everyday life (themes I had readily – perhaps too readily – associated with this 'vulnerable' group). However, following this initial encounter, the development worker suggested that I come to a few of the group meetings, to see what the organization was about before embarking on interviews around these themes. In fact, the organization itself was embarking on its own research project about links between the environment and mental health, and so it was thought my presence – as a geographer – might aid in that. (The project did happen on a small scale, resulting in a one-day workshop in which I was centrally involved.) From the outset some of my assumptions were challenged, and the nature of my research redefined. I was surprised to learn that this 'vulnerable' group was so organized, and that the people involved were themselves engaged in research, and that there could be the possibility of a very real 'intersubjective encounter' between myself and the group where 'the relationship with the researched may be reciprocal' (England 1994: 82). It rapidly became clear that, if I abandoned my quest for interviews, and instead merely 'hung out' at the offices of the group, I might learn much about how people with mental-health problems organized themselves collectively from the 'inside', and more about the diverse everyday social geographies of deinstitutionalized patients (that is, it was not all a straightforward story of 'marginality' and 'exclusion'). At the same time, it appeared that it was possible for me to help out in the formation and undertaking of grassroots research.

I began to attend the group meetings, and was welcomed as a new, if temporary participant. New members were common and so this aided my access and 'positioning'. My nervousness matched that of others (there were two new members when I came), and, as we shook and stumbled over our words in these public forums, I realized that the support evidenced for myself and other 'new people' was revealing of how the organization in part functioned as an inclusionary space for those who were inarticulate, vulnerable and marginalized. Reflecting on my own experience of access, then, told me something about the group as a whole. From the beginning my position vis-à-vis the group was one of 'student'; the organization was accustomed to accepting social-work students for placement from the local university, and I was treated very much like them (although these students were limited to six- or twelve-week placements). This ascribed position was advantageous, as the people who worked in the organization felt that I had to learn about all aspects of their work.

As I increasingly participated in meetings, I forgot about 'doing interviews' and began to immerse myself within this organization, its projects, its office politics and its social functions. Doing work for and with the organization was important in assimilating myself within it: in the beginning this involved making tea, and in the end it involved me sitting on steering groups, taking minutes, writing reports and helping to organize the day workshop. Hanging about in an office might not be considered the most immediately obvious site for a project on people with mental-health problems, but the meetings held there drew news from the discrete sites of the patient organization around the city (and facilitated my access to these). I soon recognized that I had accidentally landed in a key site for ethnographic work on this group. As I began to immerse myself within any activity I could, my 'student' status began to lessen, since, unlike the placement students, I had no official or defined role or mentor, and I gradually became a more accepted 'student-volunteer' who happened to be doing research. Finding useful material for the production of geographic knowledge involved spending long time periods (spread over the three years of the research project and beyond) accessing the office several days a week, building up relationships, stores of information and trust from other members of the patient council. There were two important dimensions to this ethnographic research, and the geographical knowledge I was involved in gathering and producing, and these dimensions were ones of which I only gradually became aware.

At first, the office seemed to be just a space that facilitated the exchange of information about what was going on in other parts of the city, and for me was a space through which I could amass details, news, conversations, policy documents, minutes of meetings and opinions. I recorded my time spent in the office in a research diary (but this was always written in private), and much of my note-taking concerned other spaces, ones located away from the office. My understanding of the office as a space of informational exchange, which served to collapse the distance between other sites of the organization and to facilitate my understanding of a wider social geography of psychiatric service-users, was always held. Yet, at times, this view also served to obscure the importance of focusing back upon the office itself as a space where particular social relations were negotiated and performed and where my positionality, as a volunteer and as a researcher, was not static. Gradually, 'doing ethnography' in the office over long time periods meant that I became aware of how this space functioned as a unique sort of work environment for people with mental-health problems, and this slow realization served to change my researcher positioning and the geographical knowledge that I was producing.

In day-to-day ethnographic encounters, it became apparent to me that most users had derived some sense of personal satisfaction from being part of the organization and the office, and that their own personal identities were partially bound up in the work that they carried out. This particular office was a unique environment in which people with mental-health problems could be engaged in meaningful work, but it also offered a supportive atmosphere that provided help and counselling if individuals felt the need. The slow realization that part of my analysis could be a commentary on the working office around me (instead of it just tackling the information that I received through being there) helped me to connect up my day-to-day notes and thoughts with other geographical studies. The performance of workplace identities, as they are described in other ethnographic studies (Crang 1994; McDowell and Court 1994; McDowell 1995), is obviously relevant, although configured differently here. To be more specific, by way of illustration, Crang laments the lack of attention to what he calls the 'neglected geographies of display' (Crang 1994: 677), which partially constitute processes of identity construction in the workplace. In 'Smokey's', a theme restaurant where Crang worked as a waiter, workplace identities were constructed around efficiency, set routines and management surveillance, but were also sometimes experienced as frenetic performances:

> At the beginning of an evening shift people talked about 'getting in the mood', standing around, doing their make work, chatting and maybe half-dancing to the music; busy nights were emotional roller-coasters, full of swapped swear words at the bar dispense area, the occasional shout of frustration at the chefs or barman, the whispered obscenity when there was no tip. (Crang 1994: 685)

In my ethnographic study, the office was there as a base for a mental-health advocacy project, which meant that there were often people accessing and working in the office who possessed mental-health problems. In a way similar (but for different reasons) to

Crang's 'emotional roller-coaster', the social space of the NPCSG office was sometimes dominated by the expression of emotion or distress, or perhaps by the less obtrusive atmosphere of depression. It could be argued that, through shared 'collaborative manufacture' (Goffman, cited in Crang 1994) over long periods of time, the social space of the Nottingham Advocacy Group office was imbued with the shared understanding that emotional performances *were* acceptable and would not be classified in the same way as they would be at the medical sites that were so often the subject of user critiques. Consequently, it became possible to rethink some ethnographic encounters such as those recorded in my research diary:

> *I like coming to do my work, because I get very depressed. I need to fill my time, because I used to work full-time. I do ... think that the people here are supportive of you ... if you have any problems, like I did with the group last week, you can get counselled by the other workers. (extract from research diary, 23 Apr. 1993)*

An atmosphere of ease was generated in the office about disclosing personal feelings and difficulties, and the 'mentally ill' label was even joked about between users. Often, as I would walk into the office in the morning, I would greet individuals by asking how they were. Some would answer 'alright for a mental patient' or 'feeling a bit mad this morning' (extract from research diary, 2 July 1993). These light-hearted comments provided a space of subversion, in that the seriousness of the label was diminished by the everyday rejection of its popular meanings. However, at other points, the emotional working space became harder to deal with as a researcher:

> *Steve came in and briefly spoke to me, saying he was depressed because of loneliness. He was aggressive and made me feel uncomfortable. The obvious way in which people display their emotions here is sometimes shocking and I don't know how to react. (extract from research diary, 15 Apr. 1994)*

Not only were these sorts of encounters difficult; the fact that I began to record them meant that my researching role had begun to blur as the boundaries of what had started as 'overt' ethnography shifted. As my rethinking of the office as a social geography to be investigated in its own right developed, so did the more 'covert' nature of my ethnographic investigation in a manner not explicitly negotiated with the organization at the beginning of the project. Although most workers in the office were not aware of this, some had taken more interest in me and what I was doing, and my previous explanations of what I had wanted to do began to take on different meanings. My status as 'insider' was thus threatened (albeit light-heartedly, as is revealed in the following encounter):

USER. So you been doing any participant observation recently then?

HESTER. Oh shut up.

USER. She's not a geographer, she doesn't know about Mann's model – she's a sociologist.

HESTER. I am not a sociologist.

USER. Go on then, tell me about Mann's model.

HESTER. I don't know about that.

USER. See, she's not a geographer.

*I started to flick through some papers in the office, hoping that [the user] would leave it alone, people were taking an interest.*

USER. Look, look at her eyes, she's doing participant observation now, watch out she's going to write something down about you.

HESTER. Leave it out will you, you'll put people off talking to me.   (extract from research diary, 22 Apr. 1994)

Here the user, by joking about my role, actually highlighted the processes of ethnography. This is perhaps a sign that many other researchers and students have been through the offices of this group, and that users have grown accustomed to the vocabulary that accompanies research. There were no serious repercussions from this incident, although it served to remind me that my 'inclusiveness' in this organization was not a taken-for-granted fact, and was thus one to be continually renegotiated. My examples here serve to show that over time ethnographic study can involve changing understandings of the spaces under investigation, and also involve change in the positionality of the researcher. Assumptions about the vulnerability of people with mental-health problems as a marginal social group were also challenged. I had initially been surprised by the strong collective organization of users, but I also became aware of other geographies through which vulnerability was articulated and negotiated, by working in the group's office space. Most significantly, though, the long time period spent researching in this space meant that its meaning as a geography began to shift, as did the boundaries of my enquiry, which in turn led to a blurring of my ethnographic/researcher status. Such a shift did not imply that my aims of giving academic 'space' to the 'voices and stories' of people with mental-health problems as a marginal social group had altered, but that the way in which relevant research materials were collected and researched had. Questions could be asked about the ethical basis of my ethnographic research, but these are difficult to answer, as my rather 'blurry' ethnography evolved from an overt to a sort of covert work gradually, without deliberate intervention. This 'blurriness' and the geographical knowledge it helped me to construct can be contrasted with my next example, where my ethnographic position was much more fixed.

## 'Fleeting' ethnography: overt research on everyday life in residential care homes

In almost complete contrast to the ethnographies carried out with the patient council group, for another part of my project I wanted to understand more about the everyday situation of people who did not access the organized forums of the user collective. Whilst I had conducted covert ethnographic work on the streets, in parks and in homeless shelters in the city (Parr 1997a, b), it was also felt necessary to access residential care homes, which formed a significant part of community care arrangements, many of which were located in the inner-city area that I had chosen to study. Access to these homes was not straightforward, since many residential care homes for people with mental-health problems are privately run, although subject to inspection by the state. Consequently, the proprietors of such homes were often suspicious of my motives to visit, and only overt, day-long, prearranged visits in each site were possible. I visited 15 care homes on separate days, usually with the intention of staying a full day to talk to residents, and with the option of returning. I carried out more covert ethnographic work around the sites of the residential homes, when my appearance, bodily movement and speech were different from how they were during the visits in the care homes themselves (see Parr 1998b for commentary on such embodied ethnographic strategies). I tried to make my overt visits towards the end of my fieldwork period, so as to protect as far as possible my covert ethnographic identity (trying to carry out both types of ethnography within the same location is a strategic nightmare!).

The day visits to the care homes were problematic for several reasons. First, short visits made intersubjective interactions with residents difficult. Second, many medical staff and social workers make trips around care homes asking similar sorts of questions to me, and I was just the latest in a long line of voyeuristic inquisitors traipsing through the living space of these individuals. Residents were justifiably suspicious of both who I was and what I wanted to know; some refused to talk and acted as if I was not there, and others were slumped in chairs, seemingly oblivious to anything external to themselves. My visits to the care homes were always written up in the research diary, and taped interviews were never carried out with residents. The people that I spoke to were often very withdrawn, and seemingly suspicious of my role as a student – to have used a tape recorder would have alienated them further – and so I became quite adept at remembering in detail short bursts of conversations in order to record them later in the day.

This was undoubtedly one of the most profoundly depressing and seemingly unsuccessful aspects of my fieldwork. The impressions of the care homes that I retain are fleeting: sometimes I would stay for a day, and yet at other times I would stay only for an hour or two. The very short visits would be the result of extremely uncommunicative residents (where I could not hide behind a covert ethnographic identity) or overbearing managers who would not leave my side when I was asking questions. At one point during the fieldwork I almost cancelled all of my other visits, as I had come to dread them and thought them to be increasingly unrelated to the wider project. This feeling changed as I began to recognize some faces in the care homes from the covert ethnographies in other parts of the city, and this enabled me to piece together the mesh of semi-institutional and other places that comprised the geography of some residents of the care homes. However, in this chapter, the purpose is to think through the problematic process of accessing residential care homes in order to carry out overt ethnographic research, and I shall use some extracts from the research diaries in order to do this.

Typically, on my arrival for my visit at a care home, the door would be opened by a member of staff, and I would immediately be taken to see the manager or owner, who usually had a distinct office within the building. My presence in staff space on first entering a building made subsequent interactions with residents difficult, as I felt I had crossed a meaningful boundary (both physical and social), and that this meant I was perhaps counted more as 'staff' or at least associated with social or nursing services. Only after this was I led to the communal space, where I was only sometimes left alone with the residents. These times were always difficult, as I was intruding upon living space, and often meeting residents who had no part in making the decision about whether I could speak with them.[2] However, more difficult were the interactions with both the residents and the proprietor when these occurred in these communal spaces:

*The husband came into the lounge and spoke to me.*
PROPRIETOR. Doctor?
HESTER. No, student.
PROPRIETOR. You'll get a few talking to you [*points to individuals around the room*] [Rob] will talk to you … he won't … and you won't get much sense out of him. [*I am looking away in shock and embarrassment, I had just been speaking to these people and he is speaking about them as though they are not there and can't understand him.*] Yeah, some will give you a civil conversation, anything you need?
HESTER. No.
[*I strongly dislike this man who appears to have no communication skills with the residents. [Rob] walks over to him and affectionately pulls on his finger. He pulls away and covers his coffee cup. 'Get off' he shouts as though [Rob] was a contaminated animal. His wife laughs.*] (extract from research diary, 15 Mar. 1994)

On the whole, although conversations were halting and often awkward in the care-home visits, residents did speak at length only when there were no staff in the room. Undoubtedly I formed impressions quickly, and staff–resident relationships were almost certainly more complex in the everyday lived experience of the residential homes than I could detect. Even so, my initial impression of a care home from a day-long visit was enough for me to be reasonably confident about what I considered to be more resident-friendly settings. The extracts that I have used above show a negative experience of care homes, and also hint at the difficulties of negotiating a research identity in such places. I was spoken to as sane, as doctor, as different from the residents, and it was sometimes assumed that I shared the same view (such as that expressed by one proprietor) of residents as incapable. In some ways I felt myself to be participating in this objectification of residents by my fleeting visits into their lives: indeed, my research was not separated from the wider processes of objectification within which these people found themselves:

*Before I went to see the residents the owner showed me round the house; she showed me each person's bedroom as if to say 'these are normal bedrooms'. I felt intrusive, voyeuristic and I stood at the door, not wanting to enter and look at other people's personal belongings. I then enter the TV room; four women are sitting down.*

HESTER. Hello, I'm just visiting, I just thought I'd come in and chat.

RESIDENT. Are you moving in?

HESTER. No, just visiting.

RESIDENT. Are you a social worker?

HESTER. No, a student.

RESIDENT. A student nurse?

*[Conversation was halting and awkward, I felt patronising, intrusive, awkward and at a loss to remember what I wanted to really know about these women. In a way this was it, this TV room was their lives, it had a view on to the world of TV and a view on to Radford out of the window. Perhaps this was enough and all I should know.]*

RESIDENT. Last time there was 5 of them?

HESTER. Who?

RESIDENT. People coming round wanting to talk to us, women and men. They wanted to know if we had any complaints, but I said I'm alright ... You've got nice hair.

HESTER. Thanks.

RESIDENT. I don't know what to say to you love.

HESTER. What's it like to live here?

RESIDENT. It's nice, I've been here fifteen years.

*[The chatty woman introduced me to Rita in the corner of the room. Throughout the chat Rita had been trying to light a cigarette, but kept dropping it. She kept patting her head and then raising her arms and shouting 'yeah'. Eventually she lit the ciggie and looked round at me, 'yeah, it's alright in here' she smiled.]*

HESTER. Do you get out and about Rita?

RITA. Piss off.

*[I knew I was being hopelessly inadequate. I was going to walk out again, my life wasn't the TV room, so how dare I ask in such cheery tones, I couldn't really ever know what it was like to spend one day a week in a day centre and the rest of the week in a TV room for 15 years. I knew it was inappropriate to be there, I wasn't one of them, I was being nosy and my questions were not really of benefit to them in any way ... When I got back to where I was living, there were two calls from university offering support as they had heard my fieldwork was not going well. It wasn't going 'well' because everyone I had met wasn't spilling their guts to me and I feel guilty. Why should they? ... I could write a Ph.D. giving*

*excuses as to why my research is fragmented and patchy, but I shouldn't really because TV rooms are the reality]*   (extract from research diary, 19 Apr. 1994)

To the reader this extract may seem unnecessary, irritating and self-indulgent, speaking more about myself than the residents of the care homes. However, this extract can also be seen as conveying something of the 'disempowering' atmosphere that was common in the care homes: something from which I as a researcher was not distanced – and could not just 'observe'. By combining reflection on my (partially 'unsuccessful') attempts to understand the care homes as dwelling places, these ethnographic experiences and clumsy, emotional notes provide a basis for the construction of geographical knowledge.

My access to these settings was not generally welcomed by the residents, and none had any direct decision in my being there. This in itself helped me to understand something about residential homes as dwelling places, where the human agency of the residents as meaningful social actors was not generally recognized or given much chance to flourish. My participation in the objectification of residents as a group to study, by asking questions of them during fleeting visits, reinforced my sense of residential homes as places that were 'semi-institutional', and where traditional understandings of 'home' and private domestic space did not hold sway. Combining reflection on my access to such settings with some of the points made by residents in my visits reaffirmed this view. For example, the lack of choice and flexibility in everyday activities such as eating frustrated one resident, as he told me about a former care home where he had lived: 'they used to give us money and we could go out and buy our own food and cook it ourselves, it was great, they don't do it like that here' (extract from research diary, 15 Mar. 1994). The caring regime at the residential homes visited meant that daily decision-making processes were often taken away from the individual residents. This aided in positive experiences of 'home' being destroyed. For example, the ways in which household tasks (such as cooking) were divided between residents and staff emphasized the 'institutionalized' nature of these residential settings. Even from my short visits, it became clear that residents usually stayed in during the day, simply watching television, and that this life was more reminiscent of institutional wards than what might be termed community living.

Thinking through my accessing of residential care, paying attention to the segregation of space and the associated interactions between myself, owners and residents, as well as noting interpretations of the 'meaning' of everyday life in such places helped me to connect these thoughts to academic literatures:

> *Home has been conceptualised as having three dimensions: the physical, which relates to objects, spaces and their boundaries; the social, involving people and their relationships and interactions; and the metaphysical, which is the meaning and significance ascribed by individuals and communities to home.*   (Willcocks et al. 1987: 4)

Even from my fleeting visits, the residents in some (if not all) care homes could be seen as being unable to participate fully in either the physical, social or the metaphysical making of 'home'. However, although this section points to the ways in which even 'unsuccessful' ethnographic access can be used and interpreted in the making of geographical knowledges, it should still be noted that such ideas can only be speculative, as the making of everyday lives cannot be understood from sitting in a room for just one day. Questions also have to be asked about whether such research strategies merely serve to reinforce the vulnerability of the people who live in these places. My concern in the previous section for ways of recording and interpreting 'the voices and stories' of people with mental-health problems as a marginal social group is present here, but in a slightly different way. Here my ethnographic presence may have reinforced the objectification of residents, even though it was also informative about the micro-dynamics of residential and semi-institutional space.

## Conclusion

The examples used in this chapter are ones that show that negotiating access in different ethnographic contexts holds different implications for the formation of research relationships and the building of geographical knowledges. I began this chapter by briefly outlining the current trend for recovering and centring marginalized 'voices, stories and geographies', and the examples used in this chapter were taken from a project that aimed to do this. In the context of the particular social group under investigation – people with mental-health problems – my strategies for ethnographic interventions can be seen as appropriate, as I attempted to understand social geographies that lay beyond the asylum and hospital. Such a project helps to 'decentre the medical in medical geograph[ical]' research (Dyck 1999: 247), and alert us to other everyday spaces where marginality and vulnerability are negotiated, subverted and lived out. Yet, the aims of ethnographic work are almost always shifted in the process of doing the research, as my examples show. Honourable intentions about 'overt' and intersubjective research relations that are 'reciprocal' and 'empowering' can be disrupted, as the grounds for building geographical knowledges are reinterpreted and newly realized within the process of participating in mundane, everyday encounters. In investigating marginality, ethnographic researchers can become embroiled within (and reproduce) the very processes that they seek to critique: and, as I have shown above, my 'objectifying' and 'unsuccessful' ethnographic access has arguably become 'ultimately data, grist for the ethnographic mill, a mill that has truly grinding power' (Stacey 1988: 23, cited in England 1994: 83). Such criticisms are worthy of debate, so that more sensitive and context-appropriate research strategies can be developed that are more attentive to the intersubjective psycho-dynamics of research relations. However, I hope that I have shown that critical reflection both during and after ethnographic research on the positionality of the researcher, and on the geographical knowledges being produced, contributes greatly to our understandings of marginality and vulnerability, and of the complicated ways in which they are experienced and articulated in everyday social life.

## Notes

1. The term 'user' was the way in which the organization referred to the people who worked there, reflecting the fact that paid workers and volunteers were also 'users' of psychiatric services.

2. This is a very interesting dilemma for researchers of institutionalized people. As Valentine (1999) has noted when discussing research on children's geographies, 'vulnerable' research subjects are often not asked to give 'consent' for research to take place, but rather 'assent' to being researched. In my example, some residents refused to talk, wandered off or fell asleep – all actions that can be conceived of as a form of 'resistance'. At the very least, these actions suggest informal ways of withholding 'consent', or avoiding 'assenting' to the research process.

## Key references

Cook, I. 1998: Participant observation. In Flowerdew, R. and Martin, D. (eds), *Methods in human geography*. London: Longman, 127–50.

Crang, P. 1994: It's showtime: on the workplace geographies of display in a restaurant in southeast England. *Environment and Planning D: Society and Space* 12, 675–704.

Desjarlais, R. 1997: *Shelter blues: sanity and selfhood among the homeless.* Philadelphia: University of Pennsylvania Press.

Dordick, G. 1997: *Something left to lose: personal relations and survival among New York's homeless.* Philadelphia: Temple University Press.

Estroff, S. 1981: *Making it crazy: an ethnography of psychiatric clients in an American community.* Berkeley and Los Angeles: University of California Press.

Parr, H. 1998b: Mental health, the body and ethnography. *Area* 30(1), 28–37.

## References

Cook, I. 1998: Participant observation. In Flowerdew, R. and Martin, D. (eds), *Methods in human geography: a guide to students doing a research project*. London: Longman, 127–50.

Crang, P. 1994: It's showtime: on the workplace geographies of display in a restaurant in southeast England. *Environment and Planning D: Society and Space* 12, 675–704.

Dyck, I. 1999: Using qualitative methods in medical geography: deconstructive moments in a subdiscipline? *Professional Geographer* 51(2), 243–53.

England, K. 1994: Getting personal: reflexivity, positionality and feminist research. *Professional Geographer* 46(1), 80–9.

Foucault, M. 1967: *Madness and civilisation: a history of insanity in an age of reason.* London: Tavistock.

Giggs, J. A. 1973: The distribution of schizophrenics in Nottingham. *Transactions of the Institute of British Geographers* 59, 55–76.

Gilbert, M. 1994: The politics of location: doing feminist research 'at home'. *Professional Geographer* 46(1), 90–6.

Jackson, P. 1988: Definitions of the situation: neighbourhood change and local politics in Chicago. In Eyles, J. and Smith, D. (eds), *Qualitative methods in human geography*. Cambridge: Polity, 49–74.

Kearns, R. A. 1990: Coping and community life for people with long term mental disabilities in Auckland. Occasional Paper No. 26, Department of Geography, University of Auckland, New Zealand.

—— and Taylor, S. M. 1989: Daily life experience of people with chronic mental disabilities in Hamilton, Ontario. *Canada's Mental Health* 37(4), 1–4.

Kobayashi, A. 1994: Colouring the field: gender, race, and the politics of fieldwork. *Professional Geographer* 46(1), 73–80.

McDowell, L. 1995: Bodywork: heterosexual gender performances in city workplaces. In Bell, D. and Valentine, G. (eds), *Mapping desire: geographies of sexualities*. London: Routledge, 75–95.

—— and Court, G. 1994: Performing work: bodily representations in merchant banks. *Environment and Planning D: Society and Space* 12, 727–50.

Nast, H. 1994: Opening remarks on 'women in the field'. *Professional Geographer* 46(1), 54–66.

Parr, H. 1997a: 'Sane' and 'insane' spaces: new geographies of deinstitutionalisation. Unpublished Ph.D. thesis, University of Wales, Lampeter.

—— 1997b: Mental health, public space and the city: questions of individual and collective access. *Environment and Planning D: Society and Space* 14(4), 435–54.

—— 1998a: The politics of methodology in 'post-medical geography': mental health research and the interview. *Health and Place* 4(4), 341–53.

—— 1998b: Mental health, the body and ethnography. *Area* 30(1), 28–37.

Philo, C. 1997: Across the water: reviewing geographical studies of asylums and other mental health facilities. *Health and Place* 3(2), 73–89.

Smith, C. J. 1981: Residential proximity and community acceptance of the mentally ill. *Journal of Operational Psychiatry* 12, 2–12.

Smith, S. 1988: Constructing local knowledge: the analysis of self in everyday life. In Eyles, J. and Smith, D. (eds), *Qualitative methods in human geography*. Cambridge: Polity, 17–38.

Staeheli, L. and Lawson, V. 1994: A discussion of women in the field: the politics of feminist fieldwork. *Professional Geographer* 46(1), 96–102.

Valentine, G. 1999: Being seen and heard? The ethical complexities of working with children and young people at home and at school. *Ethics, Place and Environment* 2, 141–55.

WGSG 1997: *Feminist geographies: explorations in diversity and difference.* London: Longman.

Willcocks, D., Peace, S. and Kellaher, L. 1987: *Private lives in public places: a research based critique of residential life in local authority old people's homes.* London: Tavistock.

# PART 5

## Interpretative strategies

# 13
# Making sense of qualitative data
## Peter Jackson

## Introduction

The aim of this chapter is to provide an example of the interpretation of qualitative data, highlighting some key issues that commonly arise. Previous chapters have discussed the process of data collection using a variety of methods, from interviewing and group discussions to ethnography and participant observation. Many of these methods involve the collection of qualitative data such as tape-recorded conversations, which are subsequently transcribed, either in full or in part. While there is plenty of advice for researchers on how to *collect* qualitative data and a growing literature on procedures for *coding* such material, there is relatively little discussion about how to *interpret* qualitative material: how to build an analysis from the transcripts of interviews or focus groups (for some exceptions, see Strauss 1987; Silverman 1993, 1997; Denzin and Lincoln 1998; Strauss and Corbin 1999). This chapter attempts to place the more technical aspects of coding and data analysis within a broader interpretative framework.

The examples in this chapter are drawn from a project undertaken in collaboration with two colleagues at the University of Sheffield: Nick Stevenson, a lecturer in sociology, and Kate Brooks, a researcher with a background in cultural and media studies.[1] The project, which lasted for just over one year, was designed to investigate recent changes in the magazine market in the UK. In particular, our project was designed to explore the commercial success and cultural significance of the new generation of men's 'lifestyle' magazines (magazines such as *Arena*, *GQ* and *Men's Health*, as well as more popular titles such as *Loaded*, *FHM* and *Maxim* (see Fig. 13.1)).

The project sought to explore the production, content and readership of the magazines, and involved interviews with editorial staff, various forms of content analysis and focus groups with a selection of readers and non-readers (see Box 13.1).[2] The interviews and focus groups generated around 30 hours of tape recording, which, when transcribed, amounted to over 700 pages of data. Even with three researchers involved in the interpretation, this represented a substantial body of material, particularly since we intended to code the transcripts manually rather than using a computer package such as NUD*IST or Ethnograph.[3] This chapter discusses some possible ways of making sense of all this material, highlighting key issues and general principles rather than providing a detailed step-by-step guide.

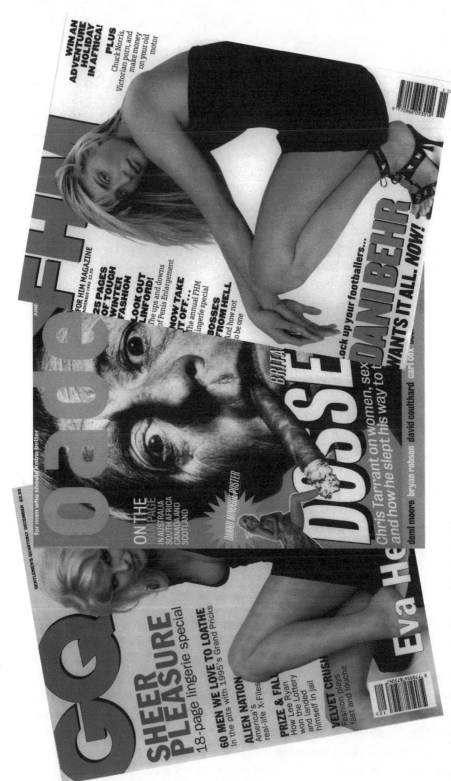

**Fig. 13.1** *Magazine covers from three of the leading titles*

© GQ/Conde Nast Publications Ltd

## Box 13.1  The focus groups

*Islington*: public-sector professionals, Asian/Jewish Londoners

*Derby women*: thirty-something, middle class, public sector

*18-year-old 'lads'*: unskilled working-class/ unemployed Sheffielders

*Bristol students*: twenty-something media studies students

*Gay men*: forty-something Sheffielders

*Bristol lecturers*: middle-class academics

*Pimlico*: 25–35-year-old casual readers, London graduates

*Musicians and artists*: middle class, Bristol

*Journalists*: professionals and students, Londoners (one man, the rest women)

*Fashion shop assistants*, working-class Sheffielders

*Turnpike Lane*: 25–40-year-old London graduates

*Stoke Newington*: media professionals, London, upper middle class

*Disabled men*: working-class Sheffielders

*Unemployed men* (with care worker): London

*Footballers*: Sheffield postgraduates and professionals

*Lecturers*: middle class, Manchester

*Art College students*, London

*Bikers* (motorcyclists): working class, thirty-something, Taunton

*Derby men*: middle-class professionals

*Counsellors*, thirty-something, middle class, London

It is important to realize that even top-quality data are of little use until they have been transcribed, coded and analysed. Even the most experienced researchers frequently underestimate the time involved in transcribing, coding and interpreting their data, leaving insufficient time for analysis and writing up. Typically, transcription requires six or seven hours for each one-hour interview (longer for group discussions where several participants may be talking simultaneously and longer still if the researcher is not an experienced transcriber).[4] Once the transcription process is complete, researchers are often faced with a mountain of data, piles of transcripts amounting to hundreds if not thousands of pages that may simply appear overwhelming. The process of extracting meaning from the data in a relatively systematic fashion usually involves some form of 'coding'.

Coding refers to the preliminary stages in qualitative analysis where the data are carefully combed through by the researcher, marking up the transcripts with a series of 'codes' that label particular words and phrases for subsequent analysis (Seale and Kelly 1998). Most systems of coding involve highlighting individual words or phrases or annotating sections of the transcript with interpretive codes involving varying degrees of abstraction from participants' own words. Anselm Strauss (1987: 33–4) distinguishes between 'in vivo' codes (terms used by informants themselves that the researcher regards as significant) and more analytical or sociologically 'constructed' codes (abstracted from the data by the researcher).[5] In our focus-group research, discussion of the magazines in terms of 'harmless fun' was an example of an 'in vivo' code (as in 'it's not very harmful … it's just a bit of fun'); terms such as 'distancing' or 'denial' were examples of more analytical codes, which were defined by the researchers and used to identify particular sections of the text.

Coding is intended to make the analysis more systematic and to build up an interpretation through a series of stages, avoiding the temptation of jumping to premature conclusions. It also encourages a thorough analysis of the transcripts, avoiding the charge that qualitative researchers have simply selected a few unrepresentative quotes to support their initial prejudices (sometimes referred to disparagingly as 'cherry picking'). While coding is rarely undertaken in a quantitative manner, in order to produce estimates of the frequency of particular codes among an identifiable 'sample' of participants, some assurances are needed regarding the reliability of the interpretation, with the process of analysis made as explicit as possible.[6] As a result, researchers are becoming increasingly self-conscious about the need to demonstrate the validity of their methods, proposing various criteria for enhancing the rigour of qualitative research and ensuring that inferences are meaningful (see Baxter and Eyles 1997). Above all, it is vital that researchers provide full information about how the data were collected, coded and interpreted. All too frequently, the actual process of interpretation remains opaque, with vague references to key themes having simply 'emerged' from the data.

## Making sense of the data

How, then, did we set about 'making sense' of our interview and focus-group material? In practice, all three researchers met on a regular basis to discuss the transcripts, taking them in batches of three or four, coding them independently and then comparing our analyses. Having several people coding independently had the advantage of providing a check on the enthusiasms and excesses of any individual interpretation. My own approach, for example, was to highlight the key words used by participants and to note in the margin more abstract, higher-level analytical codes. I found it hard to follow the advice of Strauss and others that higher-level coding should be left until a later stage in the process – a reservation that is also shared by Cook and Crang (1995). In practice, it is almost impossible to read a transcript without simultaneously reflecting on the theoretical premises or conceptual issues that led one to undertake the research in the first place.

Experienced researchers also bring to each project the 'memory' of previous research, where certain styles of analysis proved effective or otherwise. For example, in my previous project on two North London shopping centres (Miller et al. 1998) it had been hard to recruit an all-male group to discuss their shopping practices. By contrast, the ease with which men were recruited to discuss their magazine reading habits was important for our interpretation of their gendered identities. Similarly, Kate Brooks's previous research on the audience of *Judge Dredd* (the comic book and film) had emphasized the range of ways that such material is actually read (Barker and Brooks 1998). This led to an expectation that different readers would have different levels of 'investment' in the magazines, which might be expressed in terms of different 'discursive dispositions' (celebrating the magazines' humorous appeal or expressing a sense of disapproval or detachment from them). Such prior expectations do not, in my view, 'contaminate' the research. It is an inevitable feature of each successive project and can enhance the analysis, provided that researchers are reasonably self-conscious of their own agendas and preconceptions.[7]

As well as (re)reading the transcripts, it was also important to (re)listen to the tapes for those nuances of emphasis, hesitation and inflection that may not have been noted on the transcript.[8] Where a statement is made in an ironic tone, for example ('talking in italics', as one of our respondents put it), it can mean the exact opposite of its literal meaning. Working from the transcripts alone can, therefore, be quite misleading. Notes were also taken regarding the participants' non-verbal behaviour that would not be recorded on tape. This included the way focus-group participants looked at the magazines we provided as a 'prompt' to discussion at the beginning of each session, displaying examples of different 'upmarket' and 'downmarket' titles, for example. It also included the relative willingness of different group members to participate in the discussion.

At the end of each focus group we circulated a brief questionnaire, designed to solicit basic socio-economic information on each of the participants. This helped in the subsequent process of attributing each individual's contribution to the discussion (not readily done from the tapes alone, even where the precaution has been taken of asking group members to introduce themselves at the beginning of the tape). The questionnaire also allowed us to gain the written consent of participants to use their contributions in conference presentations and subsequent publications, a process that has been overlooked in much previous qualitative research but which is increasingly regarded as necessary under UK copyright and data protection law as well as being good social-science practice.[9] This is particularly the case where qualitative data are made available for public access through archiving, a point that should be borne in mind from the outset of the research.

Once the transcripts had been coded in a detailed, line-by-line, case-by-case, manner, it was then possible to provide summaries of each interview or focus group. An example of this level of analysis is given in Box 13.2. The extract illustrates some of the problems of coding. For example, summarizing individual quotations in this way removes them from the group context in which each statement was originally made. It may well be necessary, at a later stage in the analysis, to go back to the original transcript to reproduce the 'flow' of a particular conversation, to locate the context in which a particular view was

---

### Box 13.2 Summarizing an individual transcript

*The following is an extract of a transcript summary relating to one of the focus groups conducted in North London. The code is given first, followed by one or more quotations from the transcript relating to each code (with the page number from the original transcript).*

*surface/depth*

'Just not enough interest for me ... too shallow ... lacking in depth' (page 1)

*pretentiousness*

*Loaded* has 'no pretensions'; the other magazines are 'a nice version of the top shelf' (page 4); they 'attempt to be more sophisticated' (page 5); *Loaded* is 'more blatant' (page 4)

*irony*

'the flavour is probably fairly sending itself up as you read' (page 12)

*permission to be yourself*

'I think a lot of men, including myself, are asking themselves "what the fuck does it mean to be a man?" ... these magazines actually give you permission ... to be the man I want to be ... You know, whether I want to start screwing around or whatever ... It's okay to be this, this is actually who I am' (page 14).

*etc.*

---

expressed (in relation to what was asked by the moderator, for example) and to see what kind of response the remark generated among other members of the group (general consensus, laughter or violent disagreement, for example).

The extract in Box 13.2 also illustrates some more mundane, but no less significant, aspects of the coding process. Note, for example, that none of the quotations is attributed to specific participants, only to the group in which the quotation occurred. To recover the identity of specific participants involves returning to the transcript or to the original tape (where the practice of including a page or line number can be extremely useful in tracing particular quotations). Note also that, even at this stage, the transcript has been 'cleaned up' in various ways: ellipses [...] have been introduced where material has been left out; some passages are summarized rather than reported verbatim; none of the moderator's questions or prompts has been included in this example (though they are included elsewhere); and some 'tidying-up' has occurred (eliminating ums and ers, though such pauses and hesitation can be highly significant). Other 'technical' aspects of the coding process are also worth noting. For example, the second quotation, coded here in terms of 'pretentiousness', could equally have been coded as 'blatant' or 'top shelf'. A single summary of the material, such as this, is therefore only an initial step towards a more detailed interpretation of the material.

In this case, the interpretation was enriched by the involvement of my colleagues, each of whom coded the transcripts in significantly different ways. We were then able to thrash out our differences and compare interpretations. For example, I tended to summarize each transcript in turn (as above), grouping quotations according to a dozen or so 'meta-codes', at a higher level of abstraction than the participants' own ('in vivo') codes but still staying relatively close to their actual words and phrases. Kate Brooks was much more alive to the internal dynamics of the groups (no doubt reflecting the fact that she was the only one of the research team who had been present at all of the groups). She was also much more systematic than either Nick Stevenson or myself in noting the way that interviewees and focus-group participants positioned us as researchers.

This might, for example, involve noting an individual word or phrase such as 'I know it's dreadful to be politically correct' or 'we need to know how you lot think [referring to the female moderator]'. It might involve a self-conscious monitoring of sexist language by an all-male group, aware of the moderator's gender ('a pretty bird on the cover, I mean "woman", sorry').[10] Or it might involve a more general sense of the participants' evaluation of us as researchers. Examples include the editor of *Maxim*, who expected us to criticize the magazine for pursuing commercial ends rather than having a more explicit 'political' agenda, or a focus group's condescension towards the researchers as 'sad' academics who were incapable of seeing the humour in the magazines ('I think we're being far too theoretical ... I mean buying *Loaded* is not a political act in any shape or form'). Other participants placed us differently as 'right on new men' who would clearly disapprove of the magazines or as 'University people' who would look down on anyone who confessed to enjoying reading magazines. Identifying the range of positions that participants might take in relation to us as researchers encouraged us to see ourselves as part of the analysis rather than as standing somehow 'outside' or 'above' the project.[11]

Kate also challenged Nick Stevenson and myself to be as explicit as we could about our own assumptions.[12] On the basis of a position paper I had written near the beginning of the project, Kate suggested that our approach might be characterized as challenging the widely held notion of 'masculinity in crisis'; that the magazines represented a potential space for new forms of masculinity to be developed; and that the magazines had a contradictory appeal, read for entertainment and/or as a lifestyle guide. Being explicit about our own assumptions was particularly helpful in preventing premature 'closure' of the analysis or simply using the data selectively to support our own preconceptions.

Nick Stevenson's initial coding of the transcripts was more 'sociological', at a higher level of abstraction than either Kate's or mine. For example, he sought to distinguish various models of the self: a *superficial, manipulated self* ('I'll tell you, if you find a man, you know, baring his soul, talking about the interior man, you know, this is all exterior man, it's all behaviour and public face rather than private face'); an *unstable self* ('you need a model, a role, a traditional male role-model, you know'); a *false self* ('that to me is false. That to me is someone who is having to go with the flow, you know. They're not too sure who they really are, so they've picked up this image and they'll go and pay £500 just to have this particular jacket'); an *omnipotent self* ('these people are gods, these people

## Box 13.3 Comparing the transcripts

*This extract compares the transcripts from four separate focus groups, drawing out common themes and noting potentially significant contrasts between the groups.*

### Reading and buying

*'If they're in the house I'll have a look'*

The most significant thing about the question 'Why do you buy men's magazines?' is that it's usually answered by a joke (*'Damn fine articles'*: shop assistants group), embarrassed laughter or a disclaimer (*'I don't buy them, I read my friends'*: virtually everybody apart from the notable exception of the working-class group of *Loaded* readers). Why? Do women readers talk more about the mag as a treat, regularly buying, etc.? Here, this is nearly always talked about dismissively (even the Turnpike Lane subscriber was given it by his girlfriend . . .

Most groups mention they have always read women's mags, e.g. girlfriends, mothers, housemates, so this idea that men don't have the same magazine reading traditions as women is maybe not so true . . .

Men have read mags for years so reading them isn't the problem, it's buying them . . . Is it to do with men's so-called fear of commitment? (*'I'm a subscriber, but, well, my girlfriend wanted us to subscribe, I'm not that bothered', 'I just pick them up and then drop them . . .'*).

### *Loaded*

*'Loaded takes the lads' side . . . takes the mickey'*

For the *Loaded* readers we've talked to, this problem of not wanting to be seen buying/buying into the mags is resolved by the magazine itself not taking mag buying/reading seriously.

Reading *Loaded* is a laugh, this is emphasized: *'harmless fun', 'more of a laugh', 'don't take it seriously'*, etc.

etc.

have never been rejected, don't have to work for a living, they don't have normal human emotions'); and an *autonomous self* ('more single people, you know, on the market or whatever at the moment . . . If you're living on your own now for any length of time, you've got to learn to do that stuff, so it's going to change you as a person, make you more independent maybe'). Similarly, Nick sought to draw out issues of class and social distinction, including a telling reference by one of the participants to a new form of 'classless laddism'.

Once we had read and summarized each transcript in turn, comparing our independent interpretations of the data, it was then possible to attempt a second reading, *across* the interviews and focus groups rather than *within* each transcript, characterizing each group or interview according to the prevalence or absence of specific themes. An example of this stage in the analysis is shown in Box 13.3, taken from Kate Brooks's notes on four of the transcripts. This additional step in the interpretation encouraged us to see what themes were common among all our participants and which themes related to each specific interviewee or focus group. For example, some themes (such as a tendency to regard the magazines as superficial) were more common among some groups (such as the group of gay men in Sheffield) than others. Similarly, it was the groups with greatest cultural capital (for example, the group of London graduates, the Manchester Politics lecturers, the Derby public-sector women, the Bristol-based group of musicians and artists) that were most willing to discuss the question of irony. Some groups (for example, the Bristol students, the Sheffield shop assistants) clearly saw themselves as part of the magazines' target audience; others (for example, the disabled group in Sheffield, the Derby women, the Taunton bikers) defined themselves as outside the discourse used by the magazines to address their readers. Some groups (for example, the Sheffield footballers) denied any political significance to the magazines, evaluating their success in purely commercial terms, while others (for example, Islington professionals) thought the magazines could be interpreted as a 'backlash' to feminism and 'political correctness'.

The point about the relative prevalence of different themes in different groups raises a more general one: that the interpretation of qualitative material should pay as much attention to silences, absences and exclusions as to the manifest content of what is actually said. In our own transcripts there were numerous glaring absences: fatherhood was rarely mentioned; feminism occurred only in a highly caricatured form (as extreme- or ultra-feminism); 'race' was completely invisible to most of our participants; and discourses of friendship, domesticity and commitment were virtually absent (except as subjects for satire). As several commentators have noted, such silences are characteristic of the more hegemonic forms of masculinity (see Rutherford 1992; Connell 1995). Where the magazines touched on such 'sensitive' topics as men's sexuality, emotional and personal health, fashion consciousness or bodily appearance, a variety of textual strategies were used to avoid giving the appearance of taking things too seriously. These strategies included the use of 'laddish' humour, irony and other devices. The transcripts of our focus groups illustrate a range of similar strategies, returning the conversation to 'safer' ground as soon as more intimate topics are opened up (see also Stevenson et al. 2000).

The next step in the interpretation was probably the most critical to the project's success. From an initial list of 16 'meta-themes', identified from the transcript-by-transcript analysis discussed above, we spent many hours grouping and regrouping the themes, eventually highlighting 10 *discursive repertoires* that we felt most accurately characterized the way different groups attempted to 'make sense' of the magazines (see Box 13.4). These discursive repertoires are similar to Stanley Fish's concept of 'repertoire' (1980: 320), which he defined as a way of making sense of the world through systems of intelligibility that are shared by members of the same interpretive community.

> ## Box 13.4 Discursive repertoires
>
> | | |
> |---|---|
> | harmless fun | laddishness |
> | seriousness/commitment | trash/disposability |
> | honesty | irony |
> | visibility/openness | change/backlash |
> | insecurity | women as Other |

In undertaking our analysis, we were also influenced by Joke Hermes's recent work on women's magazines, where she identified a similar set of interpretive repertoires, defined as 'the cultural resources that everyday speakers … use' (1995: 145) in making sense of what they read. While these repertoires are 'not available at the level of everyday talk', they are 'the researcher's reconstruction of the cultural resources that everyday speakers may use (dependent upon their cultural capital and, thus, the range of repertoires they are familiar with) … that can be recognized as collective themes and understandings' (1995: 145).

As with most such analyses there are problems in validating the selection of repertoires: would another researcher, from a different theoretical perspective or with different political views, identify a different set of themes and, if so, would this invalidate the analysis? While Hermes clearly states her own position as a feminist, she is much less explicit about her methodology (in terms of the questions she asked her informants, for example). She also fails to make clear that her 'data' are what women *say* about the way they read magazines (at the level of public discourse) rather than what the women actually *do* with the magazines (in terms of their private reading practices).

We do not claim that the list of discursive repertoires in Box 13.4 is definitive or exhaustive of the many readings that can be inferred from our interview and focus-group transcripts. But they do allow us to make a coherent analysis of the material in relation to our initial concerns about the relationship between consumption and identity, changing masculinities and commercial cultures (for further details, see Jackson *et al.* 1999). Many other readings of our transcripts are possible. At an earlier stage in the analysis, for example, I attempted to construct an interpretation based on the distinction between 'surface' and 'depth'. As previously argued, the magazines appeared to be highly superficial in their treatment of 'serious' issues concerning men's health, sexual relationships and so on, corresponding to the glossy surface appeal of the magazines and their concern for an appropriately alluring cover. While many of the discursive repertoires identified in Box 13.4 could be interpreted in these terms, the distinction between surface and depth eventually proved unhelpful, forcing the analysis to adopt a moralistic tone in favour of fundamental change (in gender relations, for example) as opposed to the 'superficial' appeal of the magazines that many readers obviously found attractive. The surface/depth distinction also encouraged a search for 'authenticity' (in terms of genuine versus superficial change or appearance versus reality) rather than an exploration of 'horizontal' readings (tracing displacements and juxtapositions, connections and contradictions).[13]

We did not want to leave our analysis at this point, however. Having identified a range of discursive repertoires that different groups of men drew on in 'making sense' of the magazines and associated changes in contemporary masculinities, we also wanted to explore how different individuals and groups might relate to these discourses. The danger of not taking this additional step is that every reference to a particular discourse (of 'harmless fun', for example) might have been read as an endorsement of that position rather than, for example, as more ambivalent if not an outright repudiation. This is particularly important where one is concerned to explore the possibility of 'reading against the grain', suggesting oppositional standpoints or active resistance to the majority point of view.

We therefore returned to the transcripts and read them again, this time aiming to identify the range of *dispositions* that different individuals and groups took towards the discursive repertoires that we had earlier identified (see Box 13.5). While the predominant tone of the focus-group transcripts was 'celebratory' of the magazines' approach to masculinity, with their emphasis on laddish excess and sexist humour, some participants were directly hostile, others more ambivalent, adopting defensive, distanced or ironic dispositions. It is the latter that were of most interest to us in advancing our argument about the contradictory ways in which the magazines are read, signalling at least some men's ambivalence towards hegemonic representations of masculinity.[14]

**Box 13.5: Discursive dispositions**

| | |
|---|---|
| analytical | compliant |
| defensive | distancing |
| refusing/rejecting | hostile |
| celebratory | ironic |
| apologetic | deferential |
| vulnerable | dismissive |

Our interpretation therefore culminated in the identification of a series of discursive repertoires and a range of dispositions towards those repertoires. So, for example, some readers took an 'ironic' disposition towards the discourse of 'harmless fun', while others adopted a 'distanced' attitude towards the magazines' alleged 'laddishness'. The final step was to map these discourses and dispositions onto different groups of readers, trying to establish whether there were significant variations in the construction and expression of masculinity according to age, ethnicity, occupation or geographical locale (see Jackson 1991).

## Conclusion

Using some illustrations from a research project on the consumption of men's 'lifestyle' magazines, this chapter has sought to demystify the process of interpreting qualitative data. Rather than providing a step-by-step guide, the chapter has sought to highlight a number of key issues that arise at various stages in the process of interpretation. These have included:

- the tendency to underestimate the amount of time needed to transcribe, code and analyse qualitative data;

- the need to distinguish between different levels of coding, at various degrees of abstraction (from the language used by participants themselves to more analytical codes introduced by the researcher);

- the need to avoid 'cherry picking' (selective quotation) by being thorough and systematic in coding and interpreting the data, whether or not the process is undertaken manually or via a computer package;

- the need to state clearly how the process of interpretation was conducted (avoiding the suggestion that themes 'emerged' from the data or jumping prematurely to conclusions that support the researcher's initial prejudices);

- the need to include a discussion of silences, absences and exclusions from the transcripts as well as an analysis of their manifest content, and to explore the significance of humour, hesitation and non-verbal cues, which can be vital to understanding the nuances of what is actually said;

- the need to consider the researcher's positionality in relation to the research participants as an integral part of the research process;

- the relative benefits of individual versus group research (where, it was argued, the latter has some definite advantages in terms of bringing different perspectives to bear on the same material);

- the possibility of reading each interview or focus group in turn before reading 'across' the transcripts, looking for similarities and differences, connections and contradictions, among the various participants;

- the need to acknowledge that many different readings can coexist (identified here in terms of an initial distinction between 'surface' and 'depth' and subsequently in terms of discourses and dispositions) and, concomitantly, the impossibility of a single, definitive interpretation;

- and, finally, the fundamental need to be explicit about the methods used to interpret the data.

As this last point suggests, the interpretation of qualitative data is always limited by the choice of methods employed to collect the data and by the researcher's theoretical orientation (see also Ley and Mountz, this volume). For example, our research has emphasized the agency of consumers in actively shaping the meanings they derive from magazines. But our data refer only to participants' verbal representation of their reading practices (what they *say* rather than what they actually *do*). To understand their actual reading practices would require a different (more ethnographic) method, probing the way these practices, as well as the discourses on which they draw in 'making sense' of the magazines, are embedded in other aspects of their lives. Hopefully, however, this chapter has highlighted some of the issues and principles of qualitative research, focusing on the interpretation of (certain sorts of) qualitative data, avoiding some common pitfalls and making the process more transparent.

## Notes

1. The project was funded by the Economic and Social Research Council (ESRC) (award number R000221838).
2. Seven editorial interviews were undertaken, lasting between 30 minutes and an hour, mostly conducted face to face, with one undertaken by phone and one by e-mail. We also completed 20 focus groups, lasting for around an hour. These were one-off meetings, convened by Kate Brooks, often with assistance from Nick Stevenson or myself. The groups were contacted via friends and friends-of-friends and via existing youth and community groups. They included single-sex groups (all men and all women) as well as some mixed groups of men and women. A full list of the focus groups is given in Box 13.1. For a discussion of the methodological implications of different ways of recruiting and conducting focus groups, see Holbrook and Jackson (1995).
3. For a discussion of the relative merits of manual versus computer-based coding strategies, see Crang et al. (1997) and Hinchliffe et al. (1997). For a practical introduction to NUD*IST, see Stroh (1996) or Gahan and Hannibal (1998).
4. Many university departments now have access to professional transcription machines, which allow researchers to listen to a recording on headphones, using a foot pedal to go through the tape in short sections, leaving their hands free to operate a word processor. Commercial transcription services are also available but are expensive (charging around £50/hour) and reduce the researcher's personal familiarity with their material.
5. Strauss (1987: ch. 3) also distinguishes between three kinds of coding: open coding (which has a 'springboard' function for later more in-depth coding), axial coding (which concentrates more intensively and concertedly around single categories) and selective coding (which focuses on categories that the researcher has defined as central to the project).
6. As the anthropologist Clyde Mitchell (1983) has argued, the language of 'sampling' and 'representativity' is not appropriate for most qualitative research, which relies on a case-study methodology characterized by logical rather than statistical forms of inference.
7. For a discussion of the complexities involved in acknowledging the effects of a researcher's positionality, see Rose (1997).

8. This was particularly important where the interview or focus group had been undertaken by another member of the research team.

9. The ESRC's Qualitative Data Archive (Qualidata) at Essex University provides some valuable guidance on the law regarding copyright, as well as useful guidelines on confidentiality and informed consent (covering the anonymization of data via pseudonyms and similar strategies). They also offer advice on appropriate repositories for qualitative data. Their website can be accessed at http://www.essex.ac.uk/qualidata.

10. The choice of a female researcher for a project on men's magazines might need some justification. Aside from her skills as an experienced researcher, Kate's gender was a significant advantage in moderating the focus groups, the majority of which consisted predominantly or entirely of men. Kate's presence served to make the participants more conscious of their own gendered identities (since masculinity is frequently an 'unmarked', taken-for-granted identity). A male researcher might simply have reproduced, however unintentionally, the shared culture of the group, including the patterns of 'laddishness' that were a central object of the research.

11. This accords with Stanley and Wise's characterization (1993: 8) of feminist researchers aiming to be on the same critical plane as their respondents.

12. It might be worth noting, for example, that we began the project with a sense of optimism about the magazines, feeling that they signalled the opening-up of new spaces for the development of politically more progressive forms of masculinity. While this may be true in some respects (concerning the increasingly public discussion of men's health, sexuality and bodily appearance, for example), the commercial success of more 'down-market' titles, such as *Loaded* and *FHM*, is generally perceived to have led to a 'dumbing-down' of the men's magazine market, leading to the depressing conclusion that 'We're all lads now' (*Guardian*, 13 July 1998).

13. For an elaboration of these points, see Crang (1996) and Jackson (1999).

14. These points are further developed in our paper on ambivalence in men's lifestyle magazines (Stevenson et al., 2000).

## Key references

Hermes, J. 1995: *Reading women's magazines.* London: Routledge.

Jackson, P., Stevenson, N. and Brooks, K. 1999: Making sense of men's lifestyle magazines. *Environment and Planning D: Society and Space* 17, 353–68.

Silverman, D. 1993: *Interpreting qualitative data: methods for analysing talk, text and interaction.* London: Sage.

Stevenson, N., Jackson, P. and Brooks, K. (2000): Ambivalence in men's lifestyle magazines. In Jackson, P., Lowe, M., Miller, D. and Mort, F. (eds), *Commercial cultures.* Oxford: Berg, 189–212.

## References

Barker, M. and Brooks, K. 1998: *Knowing audiences: Judge Dredd, its friends, fans and foes.* Luton: University of Luton Press.

Baxter, J. and Eyles, J. 1997: Evaluating qualitative research in social geography: establishing 'rigour' in interview analysis. *Transactions of the Institute of British Geographers* NS 22(4), 505–25.

Connell, R. W. 1995: *Masculinities*. Cambridge: Polity Press.

Cook, I. and Crang, M. 1995: *Doing ethnographies*. Concepts and Techniques in Modern Geography 58. Norwich: Environmental Publications.

Crang, M. A., Hudson, A. C., Reimer, S. M. and Hinchcliffe, S. J. 1997: Software for qualitative research: 1. Prospects and overview. *Environment and Planning A* 29, 771–87.

Crang, P. 1996: Displacement, consumption and identity. *Environment and Planning A* 28, 47–67.

Denzin, N. K. and Lincoln, Y. S. (eds) 1998: *Collecting and interpreting qualitative materials*. London: Sage.

Fish, S. 1980: *Is there a text in this class?* Cambridge, MA: Harvard University Press.

Gahan, C. and Hannibal, M. 1998: *Doing qualitative research using QSR NUD.IST*. London: Sage.

Hermes, J. 1995: *Reading women's magazines*. London: Routledge.

Hinchcliffe, S. J., Crang, M. A., Reimer, S. J. and Hudson, A. C. 1997: Software for qualitative research: 2. Some thoughts on 'aiding' analysis. *Environment and Planning A* 29, 1109–24.

Holbrook, B. and Jackson, P. 1996: Shopping around: focus group research in north London. *Area* 28(2), 136–42.

Jackson, P. 1991: The cultural politics of masculinity: towards a social geography. *Transactions of the Institute of British Geographers* 16, 199–213.

—— 1999: Commodity cultures: the traffic in things. *Transactions of the Institute of British Geographers* 24, 95–108.

—— Stevenson, N. and Brooks, K. 1999: Making sense of men's lifestyle magazines. *Environment and Planning D: Society and Space* 17, 353–68.

Miller, D., Jackson, P., Thrift, N., Holbrook, B. and Rowlands, M. 1998: *Shopping, place and identity*. London: Routledge.

Mitchell, J. C. 1983: Case and situation analysis. *Sociological Review* 31, 187–211.

Rose, G. 1997: Situating knowledges: positionality, reflexivities and other tactics. *Progress in Human Geography* 21(3), 305–20.

Rutherford, J. 1992: *Men's Silences*. London: Routledge.

Seale, C. and Kelly, M. 1998: Coding and analysing data. In Seale, C. (ed.), *Researching society and culture*. London: Sage, 146–63.

Silverman, D. 1993: *Interpreting qualitative data: methods for analysing talk, text and interaction*. London: Sage.

—— (ed.) 1997: *Qualitative research: theory, method and practice*. London: Sage.

Stanley, L. and Wise, S. 1993: *Breaking out again*. 2nd edn. London: Routledge.

Stevenson, N., Jackson, P. and Brooks, K. (2000): Ambivalence in men's lifestyle magazines. In Jackson, P., Lowe, M., Miller, D., and Mort, F. (eds), *Commercial cultures*. Oxford: Berg 189–212.

Strauss, A. 1987: *Qualitative analysis for social scientists*. New York: Cambridge University Press.

—— and Corbin, J. 1999: *Basics of qualitative research: techniques and procedures for developing grounded theory*. London: Sage.

Stroh, M. 1996: Qualitative data analysis using computer packages. Unpublished paper, University of Sheffield, Graduate School Social Science Division.

# 14
# Filed work: making sense of group interviews
## Mike Crang

## Introduction

It remains a truism that the practices of interpreting qualitative data remain under-reported in geography. Rarely are they discussed in final publications, leaving the impression that qualitative material magically appeared to support arguments. Given the emotional, ethical, theoretical and practical issues so often discussed about 'field methods', it is perhaps surprising that interpretation remains the dog that did not bark. This situation is changing (see Cook and Crang 1995; Baxter and Eyles 1997; Crang 1997; Bailey et al. 1999) but the relative silence is perpetuated for three reasons. First, interpretation feels like an unglamorous almost clerical process. Second, it is an area where many feel insecure and are worried that other qualitative researchers are somehow more thorough, more creative or more incisive. Without common standards and procedures there is no shield of accepted practice to hide behind. That said, the third reason is that there is a loose and de facto common approach based around the late Anselm Strauss's and Barney Glaser's idea for 'grounded theory' creation, which has spawned several textbooks (e.g. Glaser and Strauss 1967; Strauss 1987; Strauss and Corbin 1990). This is a common approach, in the sense that other forms of hermeneutic and conversation analysis have played only a minor role in geography (for a wider set of approaches, see Silverman 1993 and Feldman 1995) though saying it is an orthodoxy is a matter of contention (see Coffey et al. 1996; Lee and Fielding 1996; Crang et al. 1997). The shared approach means the method of interpretation is rarely a matter for debate and thus gets downplayed in final works. However the process is not so mechanical or well defined that a simple citation of one of these textbooks can adequately describe what was done.

In this chapter, I will talk through a broadly chronological account of an interpretative process. Based on work with local history groups, I want to give a flavour of how 'doing analysis' feels. I shall mention mechanics but I do not want to prescribe specific techniques – some things worked for me but they may not for others. I do not want to formalize a specific way of doing things; instead I want to address the insecurity of wondering whether we are 'doing it right'. I see interpretation as a creative process, as a process of fabricating plausible stories. It is not some process of testing, breaking down

> **Box 14.1  Dialogue sample**
>
> *In this section of dialogue what interested me was not whether the claims could be verified,*
> *but how they articulated an evident nostalgia for a past way of life despite its hardships.*
>
> RAY:  You can't talk about a home nowadays compared to years ago, I mean years ago.
> JIM:  Every house was your home.
> RAY:  In our house anyway it was bare boards or if not, if we were lucky, it was lino.
> MARY:  Lino.
> RAY:  Pea soup every Sunday without doubt, if we never got anything nothing else we
> got pea soup.
> JIM:  Overcoats on the bed.
> RAY:  Overcoats! Every coat! On the bed. Everything else was on the bed you're quite
> right. But the whole point in the home then.
> BERT:  Table cloth, was a newspaper.

accounts into their constituent parts or cutting away layers to get down to the truth. In the work discussed here I was trying to reconstruct world-views, to see how participants saw their localities changing, through the way they reconstructed past places and communities. So I was not looking to pick holes in their accounts but rather to build up a picture (see Box 14.1).

So I start this chapter with a word about the project – since this frames what I was trying to do in the interpretation. Secondly, I discuss how the fieldwork became material for analysis – how it was made interpretable, transcribed, organized and so forth. Then the next step of actually trying to do the interpretation I want to look at in three ways. I consider the mechanics and practices involved. I discuss the things I hoped would help, things that sometimes did and things that in retrospect had more to do with making me feel the interpretation was working than with any merits. Finally, I want to raise a few specifics about groups that affect interpretation. In the last part of the chapter I conclude with a sense of preparing to write about all of this, since I do not want to give the impression that these are really neat separable stages. It is not as though interpretation produces results or answers that are then written up. Writing, as will be seen, was another part of trying to order and think through the material. All of these processes need to be further situated within the broad ideas that have dominated cultural geography – that there are many truths and realities, and there is no one correct interpretation. So we have a process of making interpretations about others' interpretations, which for me was coupled with a desire to allow other world-views space in my text – however contradictory and difficult that process may be (Crang 1992).

## Setting the scene

Readers will probably not be over-interested in the exact work I was doing. There are few people then (or now) who may share my fascination with accounts of neighbourhood history groups about community life in Bristol. I was looking at local history societies as an example of the ways people make sense of history in particular places (other parts of the project looked at re-enactment societies, photographic archives and heritage sites). Using group discussions fitted in with how in this case remembering the past was a social undertaking, about establishing and sharing senses of the past to produce a 'social memory'. So to set the scene for the material discussed here, these were group interviews conducted with local history societies, where I organized four successive meetings each of about an hour's duration. Previous chapters have discussed 'doing' group interviews, so I offer this solely to contextualize the work discussed here. The groups ranged from four people (on a cold day with poor public transport and a bomb scare stopping traffic) to nearly 20 (when the announcement of the research project attracted attendees who had not been at previous sessions). They were held in community rooms and were taped using a Pressure Zone Modulating microphone (giving 360 degree reception with a recording range of about 12 feet) and a Sony professional walkman.

The groups were moderated by myself alongside community workers who were regular participants with the groups. I tried to supply insightful academic questions, while the community worker dealt with more of the group dynamics – which they knew about from their involvement with the groups. Using participants who knew each other meant that, when thinking about what they said, each speaker was part of a dialogue that had started long before my project and would continue afterwards. In this sense, though specially convened around specific topics, these were not 'focus groups' of anonymous strangers so much as an ethnography of already existing groups. It was in that sense a more 'naturalistic' approach to people who were discussing things anyway (Silverman 1993). The groups allowed me to study the 'sociality of memory' – that is, how shared versions of the past were constructed through dialogue (see Fig. 14.1).

## Creating paper

At the end of the meetings I had my notes, some publications and notes prepared by the groups themselves and a stack of cassette tapes. The last I had to transcribe. Transcription is generally quite dull, so staying focused and making yourself do it is not always easy. Personally I found two hours at a sitting was my limit. The bottom line, and I am not a touch typist, was that each tape took somewhere between six and 10 hours to transcribe. They took longer if the meeting had gone well and lots of people were talking, interrupting and so forth – which makes group meetings more time-consuming than one-to-one interviews. It is also pretty near essential to use a transcription player – that is, a foot-pedal-operated tape player – otherwise you will destroy the play–pause–rewind buttons on a walkman as you go back over parts to work out what was said.

It was also at this point that I began to rue my selection of venues – as background noise unerringly obscured what seemed like crucial sections of dialogue. I could not

Mike Crang

Posy Simmonds *Every picture tells a story*

**Fig. 14.1** Posy Simmonds. *Recording events and the event of recording: oral history work has to look at how respondents now wish to present themselves and how they fit their stories together, and edit past experience. The process of recollecting the past can be as significant in shaping what is said as what actually occurred. All of us tend to try and present ourselves in a favourable light when we are recorded for posterity. Reproduced with kind permission from Posy Simmonds*

know the council were going to set up a pneumatic drill outside one building, but I should have noticed that the microphone worked by using the table it sat on as a sounding board – which meant every cup of tea put down was magnified to ear-shattering proportions. To these hazards needs to be added the special problem of groups, which is working out who said what. Getting people to identify themselves early on helps, but trying to attribute a range of interjections (from short amendments or clarifications like 'you mean East' to 'no it wasn't', through to mutters of agreement or denial) is a difficult task. Now mostly it did not matter, but later on, when I was trying to see whether someone had pursued a consistent argument, this occasionally became important (see Catterall and Maclaran 1997).

Transcribing for hours means that you become intimately acquainted with the discussions; the hardest thing is to focus rather than skim over seemingly familiar passages. Some sessions surprised me – being much more informative than I had recalled; others that I had left feeling exhilarated now seemed devoid of interest. Many sections set me thinking and making notes so I could recall the ideas they suggested. (These notes are described as theoretical memos (cf. Strauss 1987).) But I had repeatedly to bring myself back to the discipline of proceeding a line at a time – paying attention to all of the discussion and not just my favourite bits. The one constant element was, sadly, my own performance. For hours on end I had to listen to my own inept questions, to be amazed respondents could make sense of them, and to plead with myself not to ask multiple questions demanding more than one answer.

Gradually the meetings were transformed into something resembling a script of a play. A spoken, improvised event had been textualized into immutable transcripts. In the process they lost much of the flavour of dialogue and life that marks group discussions. There are notations for recording performance and intonation (see e.g. Fine 1984; Silverman 1993: 115–43), but these seemed to be overkill for the level of detail I needed. Alongside the script there were now an increasing number of annotations. I had my reactions to the meeting, I had my comments as I had transcribed, all forming gloss upon gloss around the text. Rapidly then these were becoming saturated with questions, thoughts, ideas and links to other discussions.

## 'Drawing out' themes

Bailey et al. (1999) recently responded to Baxter and Eyles's pleas (1997) for more transparency in the research process, by arguing for explicit descriptions of interpretative techniques so as to develop criteria of rigour that could stand up in, not least, a policy arena. Sensibly they suggest that analysis is better thought of as systematic rather than as standardized. Here they have pointed not just to the necessity of both transparency and creativity but also to the artificial nature of a separation of objectivity and rhetoric (Bailey et al. 1999: 170; see also Coffey et al. 1996: 3). The impression from the small selection of quotations in a final paper may be that analysis is a top-down process confirming pre-existing ideas of the researcher, whereas we need ways of representing the 'messy, creative, fragmented and complex modes of reality' (Bailey et al. 1999: 172).

While geographers have been fairly happy to invoke 'grounded theory' as resolving these issues, it is important to recognize that there are divisions and inflections. The basic outline is a blend of inductive and deductive work, where we start from transcripts and through a close reading pick out themes that we then label as 'categories' or 'codes'. Interpretation is seeing how these categories relate to each other. Most notably, the two 'founders' of grounded theory, Anselm Strauss and Barney Glaser, who really provided the most influential template with their pioneering work (Glaser and Strauss 1967), later parted company with each other, having rather different inflections of the process. Strauss in subsequent works emphasized that the 'constant comparative' method of checking codes against each other was more 'verificational' than inductive (see Rennie 1998, and reply by Corbin 1998) – that is, checking that the categories used were consistent and supportable, asking how and why one category of responses differs from or echoes another. Glaser (1992) countered with an argument for allowing the 'emergence' of ideas rather than 'forcing' them, emphasizing what he called validation rather than verification. Interestingly, Glaser expressed a worry that introducing

## Box 14.2  Annotated transcript

*This annotated transcript shows the accretion of categories and codes (in the cells) as well as continued amendment of them as I returned to them later in the light of other passages.*

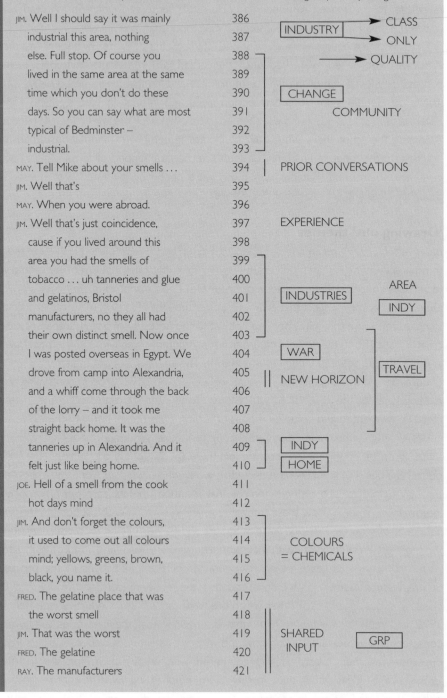

| | | |
|---|---|---|
| JIM. Well I should say it was mainly | 386 | → CLASS |
| industrial this area, nothing | 387 | INDUSTRY → ONLY |
| else. Full stop. Of course you | 388 | → QUALITY |
| lived in the same area at the same | 389 | |
| time which you don't do these | 390 | CHANGE |
| days. So you can say what are most | 391 | COMMUNITY |
| typical of Bedminster – | 392 | |
| industrial. | 393 | |
| MAY. Tell Mike about your smells … | 394 | PRIOR CONVERSATIONS |
| JIM. Well that's | 395 | |
| MAY. When you were abroad. | 396 | |
| JIM. Well that's just coincidence, | 397 | EXPERIENCE |
| cause if you lived around this | 398 | |
| area you had the smells of | 399 | |
| tobacco … uh tanneries and glue | 400 | AREA |
| and gelatinos, Bristol | 401 | INDUSTRIES |
| manufacturers, no they all had | 402 | INDY |
| their own distinct smell. Now once | 403 | |
| I was posted overseas in Egypt. We | 404 | WAR |
| drove from camp into Alexandria, | 405 | TRAVEL |
| and a whiff come through the back | 406 | NEW HORIZON |
| of the lorry – and it took me | 407 | |
| straight back home. It was the | 408 | |
| tanneries up in Alexandria. And it | 409 | INDY |
| felt just like being home. | 410 | HOME |
| JOE. Hell of a smell from the cook | 411 | |
| hot days mind | 412 | |
| JIM. And don't forget the colours, | 413 | |
| it used to come out all colours | 414 | COLOURS |
| mind; yellows, greens, brown, | 415 | = CHEMICALS |
| black, you name it. | 416 | |
| FRED. The gelatine place that was | 417 | |
| the worst smell | 418 | |
| JIM. That was the worst | 419 | SHARED |
| FRED. The gelatine | 420 | INPUT  GRP |
| RAY. The manufacturers | 421 | |

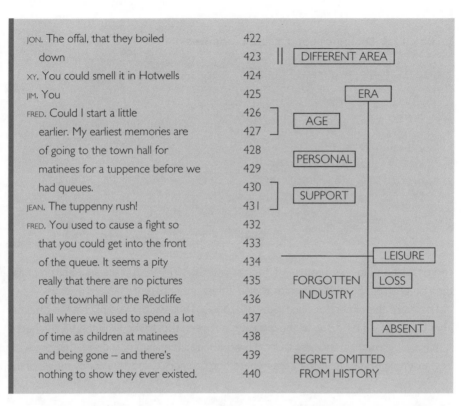

'hypothesized' relationships between categories might make the process more subjective rather than being faithful to the data. Neither 'side' could be described as following the more deconstructive treatments of texts and notions of language games – both following a cross between ontological constructivism and epistemological realism.

In this context I approached transcripts with a range of theoretical issues pressing on me – indeed my interests were marked by the questions that structure the texts. However, working through them a line at a time, slowly looking at each sentence and phrase on its own terms, ensured an openness to the unanticipated. Indeed, annotating the text with ideas and possibilities by each small section serves to remind one quickly how, in the words of Miles and Huberman (1984: 54) words are fatter than numbers with implications of meaning and context. (Box 14.2 illustrates the line-by-line accumulation of categories.) Indeed going through text like this can be exhausting – especially fighting off the urge to jump ahead with provisional ideas. LeCompte and Schensul (1999) call this 'item-level' analysis, where we are looking at individual segments and chunks of dialogue. In group discussion this may be from one speaker, but, since so much is in response to other people, it may be sequences of dialogic responses (see Box 14.3). As Lecompte and Schensul suggest, from here interpretation tends to progress to looking for patterns among the codes, at wider and wider levels. Gradually items and bits about specific topics 'pile up' and you end up with a collection of all the instances of a certain category. These categories may have subcategories, even sub-subcategories

---

### Box 14.3 Dialogic support and responses

*Group dynamics involve supporting, amending and responding to themes from other group members. Categories markings are down the right-hand margin from a computer output.*

| | | | | | |
|---|---|---|---|---|---|
| FEE. And the pounding of the mud | 338 | \| | \| | \| | −* |
| dredger and ... | 339 | \| | \| | \| | * |
| JEAN. Yes. | 340 | \| | \| | \| | * |
| BILL. The buckets ... | 341 | \| | \| | \| | * |
| DAISY. That's a past time memory for | 342 | \| | \| | \| | * |
| me cos I could lie awake and hear | 343 | \| | \| | \| | * |
| hear that ... | 344 | \| | \| | \| | * |
| BILL. .... | 345 | \| | \| | \| | * |
| DAISY. Thump, thump, thump. | 346 | \| | \| | \| | −* |

---

and/or relationships between items. Strauss and Corbin (1990) place an equal emphasis then on pushing these relationships to see if they are coherent.

These are the dimensions and axes of analysis. Thus we may simply have a category about a place and divide up all the instances between those who lived there or visited it. It may be a division, in the case of local histories, between happy and sad memories. My best example of subcategorizing was where one category called 'loss' for elements of the

---

### Box 14.4 Interpretation of transcript

*Originally all of this was under the heading of loss but here subcategories try and refine that feeling.*

| | | |
|---|---|---|
| their work ... | 1486 | SLUM |
| MAY. Jerry built. | 1487 | |
| ROGER. ... same as Ashton | 1488 | |
| HARRY. They were living close to | 1489 | INDUSTRY |
| their work with the tanneries and | 1490 | |
| Wills {yeas} ... | 1491 | |
| GEORGE. Robinsons. | 1492 | |
| MC. That's something to ask now, I | 1493 | |
| mean I come into the centre of | 1494 | |
| Bedminster all I see there is | 1495 | |
| Asda. I don't see Will's factory | 1496 | |
| or whatever. There is Asda there. | 1497 | |
| I mean how has it changed now the | 1498 | |
| industry as it were has moved out | 1499 | |
| of Bedminster if not gone | 1500 | |
| entirely. As oppose to it was | 1501 | |

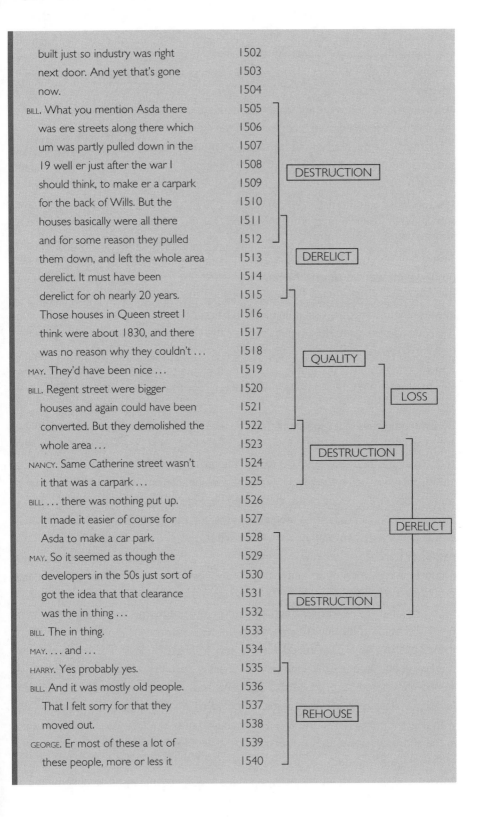

| | |
|---|---|
| built just so industry was right | 1502 |
| next door. And yet that's gone | 1503 |
| now. | 1504 |
| BILL. What you mention Asda there | 1505 |
| was ere streets along there which | 1506 |
| um was partly pulled down in the | 1507 |
| 19 well er just after the war I | 1508 |
| should think, to make er a carpark | 1509 |
| for the back of Wills. But the | 1510 |
| houses basically were all there | 1511 |
| and for some reason they pulled | 1512 |
| them down, and left the whole area | 1513 |
| derelict. It must have been | 1514 |
| derelict for oh nearly 20 years. | 1515 |
| Those houses in Queen street I | 1516 |
| think were about 1830, and there | 1517 |
| was no reason why they couldn't ... | 1518 |
| MAY. They'd have been nice ... | 1519 |
| BILL. Regent street were bigger | 1520 |
| houses and again could have been | 1521 |
| converted. But they demolished the | 1522 |
| whole area ... | 1523 |
| NANCY. Same Catherine street wasn't | 1524 |
| it that was a carpark ... | 1525 |
| BILL. ... there was nothing put up. | 1526 |
| It made it easier of course for | 1527 |
| Asda to make a car park. | 1528 |
| MAY. So it seemed as though the | 1529 |
| developers in the 50s just sort of | 1530 |
| got the idea that that clearance | 1531 |
| was the in thing ... | 1532 |
| BILL. The in thing. | 1533 |
| MAY. ... and ... | 1534 |
| HARRY. Yes probably yes. | 1535 |
| BILL. And it was mostly old people. | 1536 |
| That I felt sorry for that they | 1537 |
| moved out. | 1538 |
| GEORGE. Er most of these a lot of | 1539 |
| these people, more or less it | 1540 |

DESTRUCTION

DERELICT

QUALITY

LOSS

DESTRUCTION

DERELICT

DESTRUCTION

REHOUSE

landscape that had been lost over time became unmanageably large, and I broke it down into things that had gone entirely, things that were still going, things that were just hanging on, things of which there were fragments left and so forth (see Box 14.4). Interpretation becomes a case of looking at how patterns evolve – often through an iterative process of revising and amending codes and categories where what seemed to make sense at first later needs amending in the light of new cases.

For me the temptation was to develop ever more intricate codes and relationships, of such dazzling complexity that I could hardly remember them let alone use them. Indeed, it is a real discipline to remember each and every code and think whether it can apply to every instance – not just the first that comes to mind but to check all of them. I was working with a simple program (Ethnograph) for this project, which allowed up to 13 codes to be associated with a piece of text. Although the designer (John Seidel) was clear this did not mean you *should* have 13 categories on everything, the temptation kept pushing me. I was working by marking up the text first then entering the categories into the computer. At this stage I radically simplified the code structure from the original intentions into something usable. In part this was because manually entering line numbers and code names was so slow, but it also meant I was once more closely reading the text. Current software, where you highlight text, select a category and click (one-step on screen coding), speeds this, but it can reduce the time to think about the text (Catterall and Maclaran 1997). (For a review of software, see Crang et al. 1997; Hinchliffe et al. 1997; Barry 1998.)

Many techniques that appeared to add rigour to the analysis tended to be top down, imposing ideas on the material. Since I was responding to ideas of 'history from below' and allowing other stories to be told, I wanted to feel that I was facilitating group participants, not just in the meetings but also in the interpretation. Invoking a sense of 'authenticity' in the voices of the group members meant I eschewed certain models of 'rigorous' analysis. Equally, even though I was thinking of representing them as a 'collation' or collage of voices, my authorial role was still strong. Any idea that interpretation was passive or transparent was quickly dispelled by the graft involved. I could hardly claim the products were pristine or 'untouched'. One formal approach that did help was producing 'code maps' (see Fig. 14.2). These started as no more than my notes, where I kept a list of all my 'working' categories – so I would not forget any or overlook them. Gradually, as I kept looking at them and at the text and working around, I began linking them up; then, as a page risked turning into a tangle of lines, it might even mean redrawing the page so that links stopped being so messy. Of course, many pages stubbornly refused to produce this problem, and connections there were sparse – forming a silent reproach that perhaps I could have run that group better. This was worrying, since they lacked the complexity and richness qualitative work celebrates, but in the end there was a fuzzy but perceptible difference between useful, interesting and irrelevant sessions. 'Semiotic clustering' proved helpful in drawing out connections between categories about planners, the state and comments about local communities

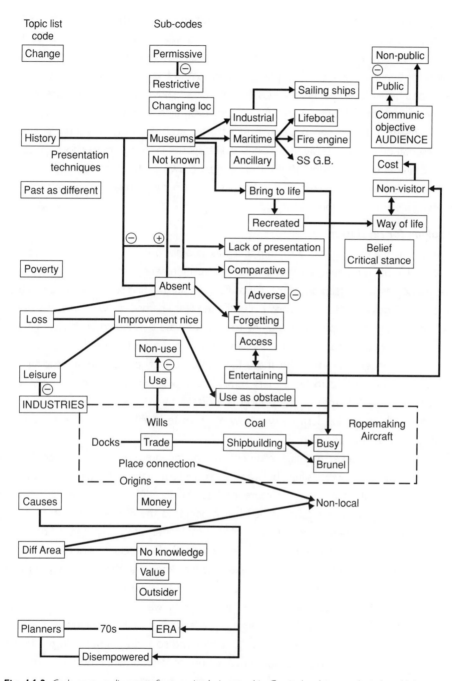

**Fig. 14.2** *Code map: a diagram of categories being used in Group 1c, drawn as I tried to think through the codes and their relationships. See at the middle, 'Industries' has been enclosed in a dotted line, as my category included mentions by group members of Ropemaking, the Docks, Aircraft, Coal and so forth. The category 'Leisure' superseded 'Leisure versus Work', written over with a 'NO' when it had to be deleted. The densest level of coding represents subthemes characterizing how local history is presented*

losing control of their own environments. Metonymy, where the part stands for the whole, was quite important in seeing that 'planners' was used as a shorthand for 'the state', and blurred across into 'anonymous corporations'. Semiotic clustering is a grand name for bringing together overlapping categories and trying to tease out if they were related to higher level 'meta-categories' (in Fig. 14.2, see how the categories generally get larger towards the left of the sheet; for finalized examples see Feldman 1995: 31). Since doing this work I have become drawn to Atlas/ti software, which allows the graphical representation and manipulation of codes and categories in a range of free form patterns (see Crang et al. 1997; Barry 1998). Links can be drawn between codes, and relations between them defined – not only subcodes but also causal relations, or relations of metonymy and opposition.

### Looking for revelation

Interpreting the material had a rhythm that was a combination of uncertainty, hope and anxiety. I kept anticipating some moment of revelation, where the 'hidden' meaning in the transcripts would leap out at me. Of course, I wanted to find something unexpected, since that would 'validate' my work. An unspoken fear I harboured was, what if there is nothing remarkable. I spent hours trying to think of ways that would enable something previously unnoticed to appear. By and large, no such moment of revelation occurred. There were indeed surprises, but these tended to be early on as I began reading and transcribing and tended to be salutary reminders about how partial my memory was, how I had missed themes or not noticed important comments from group members. The generally reassuring surprise was how much material there was in the transcripts that was informative and helpful.

There were, however, no magic moments where 'the interpretation' appeared. This is normal in qualitative work, sifting between various interpretations, building more or less stable notions of other people's more or less shifting ideas. For me there was an incremental development. Given that I had moderated the meetings, transcribed the tapes, categorized them, reread them, entered categories into computers, I knew the material intimately. So there were few surprises left. Boolean searches of combinations of categories were useful in finding examples of what I knew was there and checking that what I thought was the case was indeed so. They also gave me a chance to see whether the point was made repeatedly, by several people, or was unique. In that sense they were ways of checking a much more organic process. Most of the techniques I tried did not 'work' – so searches for sequences of codes and so forth did not help me in this project – but meant I was pretty sure I had left no stone unturned. I used diagrams and formal techniques to help me think rather than to prove the 'reliability' of my interpretations.

### Things that count

One of the ways of trying to wring coherence out of material that seems appealing with group work is content analysis. I know this appeals partly because groups have more people saying different things. Content analysis, as the quantification of qualitative data, is not something I find that attractive, but, that said, most analysis uses quantitative indicators in a loose fashion. So I have already mentioned how I went back to check how often particular categories overlapped, whether a point had been made repeatedly and so forth. This is a natural part of interpretation, but formal statistical testing is really difficult to apply. So, while it may be helpful to know if one participant or several raised an issue, that does not form a direct estimate of how important it may be. It can be very easy to produce statistics, but I have found few that really helped.

I was initially interested in looking at the percentage of occurrences where one category overlapped with another as a way of formalizing categories and subcategories, but, given the iterative process of coding, revising and developing ideas, this really did not help. There was no moment when I could turn a handle and see if something came out. It is rare for group interpretation to start with fixed (and tightly defined) codes that are applied and then the pattern interpreted. Instead the pattern is continually playing in your mind as you work, meaning I tended to go back and forth revising codes and reapplying them. They were not set up for 'reliability tests', being no more or less reliable than my interpretation. Consistency of application was certainly important, and helped by computer software retrieving segments of text, and where there were inconsistencies it suggested that the categorization or interpretation needed to evolve, but it did not 'prove' things since in the end I was more interested in what was inside categories than abstract patterns. They were a means to this end.

Some statistics were helpful in a supporting role. So looking at how often something was said did give you some ideas, but, since each occurrence could vary in length, crude counts were out. Likewise, group analysis must also look at who says what – that is, not only whether one topic came up several times but who said it. Sometimes a topic I recalled as very important turned out to be prominent because of one discussant harping on about it. Frequency counts on their own were not that useful unless linked to speakers. One salutary statistic this threw up, when the groups had been moderated, was who spoke how often. Sometimes it became clear that I had done a poor job at stopping a couple of people dominating the conversation. But equally some people who counted as contributing lots were principally offering supportive comments, while some others who spoke only rarely made (sometimes excessively) long contributions.

One of the things that I noticed losing here was the sense of force and emphasis. There is a clear difference between someone saying something quite a lot compared to saying it loudly or then again saying it definitively and retreating. These are things that in moderating a group you try and work through – getting people to elaborate, explain and so forth. But with the dry transcripts this all too often seemed to melt away. The dynamics of conversations where some set the tone and others followed seemed to me more interesting, and to say more about the topics discussed than measures of how often, or even for how long, individuals contributed.

**Fig. 14.3** *A now demolished suburban high street of the 1930s (Hotwells in Bristol) – a poignant and privileged landscape of memory for the oral history groups. The decline of these local centres in working-class areas was taken as indicative of a loss of neighbourhood cohesion in many discussions. The areas thus formed one of the focal places of remembered community life and were sites of shared stories for reminiscences*

### Fragmented stories

One criticism of interpretation based around coding, categorizing and so forth is that it works by fragmenting material. Snippets risk becoming decontextualized or rather recontextualized in new analytic frameworks. It is indeed a worry that increasingly as you get drawn into working with themes, categories and so forth you begin to see the material in terms of manipulating detached segments of text. The original flow and pattern get lost (Catterall and Maclaren 1997). This is perhaps inevitable, when even finished life-history accounts, which represent a straight narrative, are reconstructed from many question and answer exchanges in life-history interviews (Oring 1987). In other words, groups tend to start from fragments and linear stories are the exception. Some participants would indeed provide lengthy monologues about places or issues that were important to them (see Fig. 14.3), but, when groups were working well as groups, then most offerings were interrupted, supported, elaborated or amended by others.

I was especially interested in this as it seemed to me a better reflection of the praxis of social memory. What I wanted was to capture a sense of how a shared past was constructed through dialogue and discussion (Schrager 1983). I was asking whether the narrative and chronological histories that were so often recounted on the basis of lengthy life-history interviews (cf. Bruner 1987; Kirshenblatt-Gimblett 1989) could be set against fragmented, located stories about places that were derived through groups working on local history. In that sense I was using groups because I did not want to try and develop narrative interpretations (for debates on the role of narrative in British popular historiography, see e.g. Maynes 1989, 1992; Somers 1992; Steinmetz 1992). So the segmented nature of group material I found was very susceptible to being divided into chunks by topics and themes, and could then be reassembled quite effectively. However, I was not building narrative or temporal categories. I could do so at a grand scale, with time periods, likewise at a micro-scale, where explicit contrasts of 'now and then' were made by participants. Standard 'grounded-theory'-style categorizing is not that strong on getting a sense of the flow of conversation and changing opinions. As part of her assessment of different software, Barry (1998) suggests using hypertext links to trace the development of ideas to get round this.

### Paper piles

Throughout this chapter I have been conscious that discussions of 'how-to' methods have the tendency of making interpretation appear reassuringly doable yet aridly mechanistic. I am concerned not to lose sight of what, to me, was the main incentive in the process – which was to hear the views of participants. While inevitably my interpretation was an imposition on theirs, and while being aware that the idea of 'learning from the feet of the oppressed' is both an uncomfortable and improbable position (Popular Memory Group 1982), a sense of history from 'below' was important to me. That meant that in the final piece I did not want to talk about categories or abstractions to the exclusion of the voices of the local history groups. Of course, their voices would not magically 'speak through' my work, and if there was ventriloquism it was more probably me controlling what topics they 'talked about' in the staged and repeated dialogue on the pages. Nor was I trying to create a 'realist' account portraying a people and place 'as they were' (Atkinson and Coffey 1995). I was also aware that the inclusion of 'authentic voices' is a credentializing technique, and rhetorical strategy, just as much as citing references. However, even knowing this, it was important to hear things in these people's own words. The abstractions and categories were a vehicle for organizing my thoughts – not an end. Throughout the process my closeness and proximity to the data varied and I would be suspicious of anyone claiming we should be wholly detached from or submerged within it – detachment and proximity are very loaded terms in discussing the reliability of interpretation (Richards 1996). At some moments I would be more involved in thinking about categories or abstract patterns – especially as interpretation progressed – but when it came to writing then I found myself returning to the speakers, to the voices and comments. All of which can be added to the reasons why the process of interpretation tends to fade out of the final work.

Thinking through categories that developed through the process of interpretation meant that I had a fairly clear idea of themes to discuss. I produced something like an outline chapter in terms of the categories that I had developed. I then set to, working to search through categories to recall relevant material. This produced piles of material in each category that I sifted through – looking for the 'best' informant quote but also trying to give a flavour of the sort of responses. I then used these more fine-grained piles of text to work to construct the flow of the discussion within each section, scrupulously including differing views and looking at how various themes played out. My aim was to produce something like a collage from all the segments and fragments that coding had generated and from them to offer some sort of path through the material. I wanted to keep a sense of the dialogue and debate to reflect how what was produced in these groups was not a fixed and definitive history but a negotiated and evolving one, where if there was a dominant view it was reached through discussion.

This was a deeply satisfying process, as the categories really did help guide me through the way discussions had gone (see Box 14.5). This was indeed a really easy phase, as the thinking over categories and the piles of text fitted together. Writing felt like a dialogue with both the participants and my ideas, as theoretical ideas, contextual reading and so forth had informed categories that helped me organize and think through participants' words. Software allowed quoted material to be output and joined into word-processed

---

**Box 14.5 Report outline**

*Section outlines of a report to Avon Oral History Network, based on major categories, with subcategories derived from coding*

2     Decline of the Community
        Neighbourliness destroyed
        Neighbourly? When?
        Progress and Private Gains
        Erosion of Public and Communal Institutions
        A Geography of Loss

3     Almost a Revolution
        The Past versus the Present: Then, Now
        Hard to Imagine what it was like
        So much change, so fast

4     Differing Voices
        Time it changed, I wouldn't go back
        Community as control
        Poverty and charity
        Alright for the boys

5     Preservation and resistance
        We need a union!
        Who takes action
        Different areas different experiences
        An increased local input
        Ordinary Experience and Official Histories

drafts with minimum friction. It was all so easy I only belatedly noticed that the word count was rising ominously over 80,000. So the painful bit was then really honing the argument, deciding which nuances and themes could be discarded, which dialogues could be shut down. This was a much more painful process, but once again the way it worked was not through abstract schemas but through reading and thinking about each quotation, the purpose it served and so forth. In this sense the interpretative process came back to where it had started. It also suggests one of the reasons why counter-examples and different interpretations get eliminated from final papers.

The practice I have tried to sketch here is not meant to suit all possibilities. But I hope to have suggested that interpretation needs to be set within a context of what the research is trying to achieve. I have tried to hold together three somewhat contradictory impulses. First, that techniques can help both practically and psychologically in dealing with large amounts of qualitative material. Second, that they do enable some kinds of systematicity but they cannot be mechanistically applied, and patterns are emergent rather than designed. Third, that the process is set in the context of the double hermeneutic of trying to develop our interpretations of other people's contested and not always coherent interpretations and convey these more or less persuasively to others. Coffey et al. (1996) suggest these last two are opposed trends, with a semi-realist method driven by grounded theory and a pluralist experimental epistemology in other quarters. My work has tended to side with the latter, but I do not think this latter is necessarily a reason to abandon all notions of orchestrating material – as Walter Benjamin once noted, there is all the difference in the world between the presentation of confusion and confused presentation. However, it does mean that we are rarely in the business of uncovering patterns and order, more in the business of creating and making them. Systematic work does not bypass or obviate critiques of ethnographic modes of representation; it is an authority claim that makes certain realities more visible (see Atkinson 1990). I think at the end this leaves us having to think further about where we started, which is the appearance of final text. Interpretation as a practice goes all the way through to writing and ways of representing material. Thinking of how, in this case, popular memory works and relating that to methods, interpretation and representation are not separate issues.

## Key references

Atkinson, P. 1990: *The ethnographic imagination: textual constructions of reality.* London: Routledge.

Crang, M. 1997: Analysing qualitative materials. In Flowerdew, R. and Martin, D. (eds), *Methods in human geography: a guide for students doing a research project.* London: Longman, 183–96.

Feldman, M. 1995: *Strategies for interpreting qualitative data.* London: Sage.

Silverman, D. 1993: *Interpreting qualitative data: methods for analysing talk, text and interaction.* London: Sage.

Strauss, A. 1987: *Qualitative analysis for social scientists.* New York: Cambridge University Press.

# References

Atkinson, P. 1990: *The ethnographic imagination: textual constructions of reality.* London: Routledge.

—— and Coffey, A. 1995: Realism and its discontents. In Adam, B. and Allen, S. (eds), *Theorising culture.* London: UCL Press.

Bailey, C., White, C. and Pain, R. 1999: Evaluating qualitative research: dealing with the tension between 'science' and 'creativity'. *Area* 31(2), 169–78.

Barry, C. 1998: Choosing qualitative data analysis software: Atlas/ti and Nudist compared, *Sociological Research Online* 3(3), http://www.socresonline.org.uk/socresonline/3/3/4.html.

Baxter, J. and Eyles J. 1997: Evaluating qualitative research in social geography: establishing rigour in interview analysis. *Transactions of the Institute of British Geographers* NS 22(4), 505–25.

Bruner, J. 1987: Life as Narrative. *Social Research* 54(1), 11–32.

Catterall, M. and Maclaren, P. 1997: Focus group data and qualitative analysis programs: coding the moving picture as well as the snapshots. *Sociological Research Online* 2(1), <http://www.socresonline.org.uk/socresonline/2/1/6.html>.

Coffey, A., Holbrook, B. and Atkinson, P. 1996: Qualitative data analysis: technologies and representations. *Sociological Research Online* 1(1), <http://www.socresonline.org.uk/socresonline/1/1/4.html>.

Cook, I. and Crang, M. 1995: *Doing ethnographies.* Concepts and Techniques in Modern Geography 58. Norwich: Environmental Publications.

Corbin, J. 1998: Alternative interpretations: valid or not? *Theory & Psychology* 8(1), 121–8.

Crang, M. 1997: Analysing qualitative materials. In Flowerdew, R. and Martin D. (eds), *Methods in human geography: a guide for students doing a research project.* London: Longman, 183–96.

—— Hinchcliffe, S., Hudson, A. and Reimer, S. 1997: Software for qualitative research: 1. Prospectus and overview. *Environment and Planning A* 29, 771–87.

Crang, P. 1992: The politics of polyphony: reconfigurations in geographical authority. *Environment and Planning D: Society and Space* 10(5), 527–49.

Feldman, M. 1995: *Strategies for interpreting qualitative data.* London: Sage.

Fine, E. 1984: *The folklore text: from performance to print.* Bloomington: Indiana University Press.

Glaser, B. 1992: *Emergence vs. forcing: the basics of grounded theory analysis.* Mill Valley, CA: Sociology Press.

—— and Strauss, A. 1967: *The discovery of grounded theory: strategies for qualitative research.* Chicago: Aldine.

Hinchcliffe, S. J., Crang, M. A., Reimer, S. and Hudson, A. C. 1997: Software for qualitative research: 2. Some thoughts on 'aiding analysis'. *Environment and Planning A* 29, 1109–24.

Kirshenblatt-Gimblett, B. 1989: Authoring lives. *Journal of Folklore Research* 26(2), 123–49.

LeCompte, M. and Schensul, J. 1999: *Analyzing and Interpreting Ethnographic Data*. Walnut Creek, CA: AltaMira Press.

Lee, R. and Fielding, N. 1996: Qualitative data analysis: representations of a technology: a comment on Coffey, Holbrook and Atkinson. *Sociological Research Online* 1(4), <http://www.socresonline.org.uk/socresonline/1/4/lf.html>.

Maynes, M. 1989: Gender and narrative form in French and German working-class autobiographies. In Personal Narratives Group (ed.), *Interpreting women's lives: feminist theory and personal narratives*. Bloomington: Indiana University Press, 103–17.

—— 1992: Autobiography and class formation in nineteenth century Europe: methodological considerations. *Social Science History* 16(3), 517–37.

Miles, M. and Huberman, A. M. 1984: *Qualitative data analysis: a sourcebook of new methods*. 1st edn. London: Sage.

Oring, E. 1987: Generating lives: the construction of an autobiography. *Journal of Folklore Research* 24, 241–62.

Popular Memory Group 1982: Popular memory: theory, politics, method. In Johnson, R., McLennan, G., Schwarz, B. and Sutton, D. (eds), *Making histories*. Birmingham: Hutchinson/CCCS, 205–52.

Rennie, D. 1998: Grounded theory methodology: the pressing need for a coherent logic of justification. *Theory & Psychology* 8(1), 101–19.

Richards, L. 1996: Closeness to data: the changing goals of qualitative data handling. Paper delivered to the International Sociological Association, Methodology conference, Essex, UK.

Schrager, S. 1983: What is social in oral history. *International Journal of Oral History* 4, 76–98.

Silverman, D. 1993: *Interpreting qualitative data: methods for analysing talk, text and interaction*. London: Sage.

Somers, M. 1992: Narrative, narrative identity and social action: rethinking English working-class formation. *Social Science History* 16(4), 591–630.

Steinmetz, G. 1992: Reflections on the role of social narratives in working class formations: narrative theory in the social sciences. *Social Science History* 16(3), 489–516.

Strauss, A. 1987: *Qualitative analysis for social scientists*. New York: Cambridge University Press.

—— and Corbin, J. 1990: *Basics of qualitative research: grounded theory procedures and techniques*. London: Sage.

David Ley and Alison Mo

# 15
# Interpretation, representation, positionality: issues in field research in human geography
## David Ley and Alison Mountz

## Introduction: interpretations all!

Interpretation is not new to geographical research. If classical regional geography often saw its mission as one of describing the people and places on the earth's surface, those descriptions were always also interpretations, for understanding the human use of the earth was brought against the context of the natural environment and beneath the presuppositions of an Enlightenment and Eurocentric world-view. In the 1960s the sometimes unimaginative and hackneyed style of regional work was labelled by critics as 'mere description', and was replaced, ironically enough, by a frequently more sterile form of description. In the name of spatial science, places were simplified to undifferentiated and standardized surfaces upon which spatial logics unfolded, often represented through mathematical notation, while human nature was grossly reduced to a few elementary decision rules. In the 1970s humanistic geographers reacted vigorously against the theoretical, political and ethical errors in this representation of people and place, and, drawing upon sociology, anthropology and such philosophies of meaning as phenomenology and existentialism, began an exploration of the human experience of place. Together with more scholarly work in cultural/historical geography, the concern with qualitative methods and interpretative research originated in this literature.

During the 1990s, qualitative work spread into virtually every corner of the discipline, including such formerly positivist strongholds as economic geography and medical geography. While some purists would want to make this diffusion complete and argue for research entirely without measurement, a more balanced position favours methodological triangulation, moving around a research project with whatever methods offer the best solution for the problem at hand. The strategy of triangulation combines quantitative and qualitative approaches as appropriate, while recognizing the strengths and weaknesses of each.

An early objection to qualitative methods was their supposed subjectivity, depending as they do on the judgement of the researcher. However, that particular argument has been muted by work in the sociology of knowledge that made the important point that all methods are unavoidably social products. An apparently objective mathematical method like factor analysis, for example, requires a whole series of judgement calls: which variables should be *selected* for input, which statistical rotation should be *chosen* in analysis, how many factors should be *considered*, and, not least, how they should be interpreted. In a real sense, we might argue that all methods in human geography are interpretative, involving the application of a set of methodological conventions by a value-impregnated researcher.

At the same time it is important to be mindful of the need for rigour in qualitative methodology, a discussion that remains at an early stage of development in human geography (Baxter and Eyles 1997; Bailey et al. 1999). However, aside from internal concerns about rigour, an important advance has been the recognition of the ideological burden of interpretation, the depth of socialization and value-ladenness of researchers as we aim to interpret local geographies. We are all caught up in a web of contexts – class, age, gender, nationality, intellectual tradition and others – that shape our capacity to tell the story of others. Work in cultural anthropology in the 1980s outlined this 'crisis of representation', identifying how the stories told by European anthropologists about non-European cultures were uncomfortably coloured by the narrative of imperialism. Geography with its long history of exploration and map-making, often under royal or commercial patronage, shared the same ideological biases.

In human geography as elsewhere in the humanities and social sciences, the crisis of representation has caused considerable intellectual hand-wringing (Duncan and Ley 1993). Two principal issues have been at stake. First, to what extent is interpretation of the 'other' an act of social and cultural privilege, and as such an exercise in unequal relations of power? The researcher is typically articulate, well educated and socially and economically privileged, able to reach and influence a like-minded audience. The story the researcher tells becomes part of a pool of shared knowledge, and that knowledge can itself influence the actions of privileged groups towards peoples in more marginalized settings. This anxiety has raised the question for some researchers whether interpretation of the 'Other' is ethically defensible, and under what circumstances it could be so. But there is also an epistemological question. Is representation of the Other even possible, when researchers are so thoroughly saturated with the ideological baggage of their own culture? When we as researchers bring so much personal and social clutter to an interpretation, to what degree does the interpretation become, at least in part and perhaps more than in part, a construction of our own biographies? Scholars adopt varied positions on this question. Some agree with Rorty (1991), taking a strong constructionist position, arguing that ethnographers are 'tricksters', their interpretations 'fictions', a view 'more Nietzschean than realist or hermeneutic [where] all constructed truths are made possible by powerful "lies" of exclusion and rhetoric' (Clifford 1986: 6, 7). A less radical and more conventional hermeneutic position diverts the bad faith implicit in Clifford's challenge by acknowledging that the intervention of a value-laden researcher is an inevitable part of the construction of knowledge. Resolution comes, though incompletely, through an acknowledgement of the tentativeness of an interpretation, and a rigorous process of self-criticism to exorcise the demons of bias. We shall seek to illustrate these points in greater detail as we turn to our own research experience, drawing upon interpretative inner-city work in Vancouver and Philadelphia, and a transnational study situated both in Poughkeepsie, New York, and Oaxaca, Mexico. In conclusion we shall cast the net a little more broadly in assessing lessons to learn from interpretative research attentive to questions of positionality and representation.

## Inner-city research in Vancouver and Philadelphia

The objective of interpretative methodology is a nuanced story that must be retrieved through detailed knowledge of a locale and a time, gained through months or years of study (Harris 1978).[1] The task of interpretation requires building up a substantial stock of knowledge in order to frame an understanding of particular events. A moot point is the place of theory, or indeed any knowledge not gained directly from the field, in this interpretation. Experiences from both of the studies that follow have encouraged us to hold lightly onto the expectations that pre-existed field research, making them accountable to the fieldwork, rather than, as sometimes happens, imprison field data within preconceived positions. The field researcher must never suspend the capacity to be surprised.

The fieldwork in question was undertaken in two studies 20 years apart with inner-city residents with profoundly divergent life chances. In Vancouver in the early 1990s a conflict emerged between competing elites concerning the appearance of houses and streetscapes in Shaughnessy, the city's pre-eminent neighbourhood of old money (Ley 1995). An older Anglo-Canadian elite wished to see the reproduction of a traditional landscape filled with images of anglophilia: mature trees and frequently picturesque landscaping, together with a limited repertoire of house styles drawn from north-west European precedents, in which the Tudor revival held most-favoured status. The preservation and the reproduction of this landscape implied the survival and continuity of a particular subculture in the city's richest neighbourhoods. A new elite, comprising wealthy Chinese-origin immigrants, principally from Hong Kong, similarly wished to see the landscape of the same neighbourhood evolve in *their* own image. Their imagining privileged the new, the modern, requiring property demolition and (initially) clear-cutting of trees and vegetation to create a *tabula rasa* upon which a new landscape pointing to new social and economic relations might prevail. The erasure of the past was the erasure of the identity and power of those who had shaped the past; the new form of the landscape was a statement announcing a new distribution of economic and political power.

This research generated a number of practical difficulties concerning the position of the interpreter. As I had been a resident of an adjacent district for a number of years, I had acquired considerable knowledge of the research site, but my knowledge was unequal, much fuller of the old community than the new. The old community was English-speaking, while the new usually spoke Cantonese and sometimes Mandarin, and was therefore much less accessible to a uni-lingual researcher. Since the old community also shared my ethnic identification, the capacity for self-projection and a biased interpretation was considerable.[2] A significant challenge was to gain adequate access to the new community. This was secured through publicly reported statements in the media by Hong Kong immigrants, through an analysis of the letters they sent to city council, and by recording briefs they presented at public meetings, usually first in Cantonese and then in English through an interpreter. While my representation of these events was satisfactory in the light of subsequent research with this immigrant community (Ley 1999), in hindsight more could have been done, particularly in terms of household interviews in Cantonese or English, to make sense of their making sense of the conflict in which they were embroiled. At the same time it is likely that, if I had done the interviewing, my ethnic identity might have led to a rather sanitized version of the issues in conversation.

A second interpretative challenge was to understand the reasons for the conflict. The issue became highly partisan, a *cause célèbre* not only in Vancouver and across Canada but also in the Hong Kong media. Among the combatants, the land-use controversy was variously cast in terms of heritage preservation (legitimate) or landscape taste (arbitrary), of community control (legitimate) or obstructions to property rights (illegitimate), or, most tendentiously, as identity maintenance (legitimate) or racism (reprehensible). Both sides worked with the account they felt carried most political weight, and among the agents of change, particularly the advocates for a real-estate industry benefiting from land sales and property construction, the unanswerable argument was the accusation of racism against the old elite. There was in short a politics of explanation (Ley 1995) among the diverse parties. No less political have been some academic accounts of the conflict (and others like it elsewhere in Canada and the USA). In interpretation it is important to remain – as far as possible, and this is never far enough – sceptical of *all* accounts, rather than hone in on an interpretation whose politics seem congruent with the author's own. This is itself a political judgement about which there is considerable discussion and some disagreement. While value neutrality is not attainable, researchers need to weigh carefully the degree to which their interpretation, particularly in such charged areas as race, gender or class, slips too readily into a partisan discourse. Treading an interpretative path through this minefield of politically freighted accounts, where public statements could well be self-serving, and academics could easily be intended or unintended accomplices, was a significant challenge.

Issues of positionality had been encountered 20 years earlier in fieldwork concerning a much more ephemeral landscape trace, the practice of scribbling graffiti in public and private spaces (Ley and Cybriwsky 1974). This research was part of a. larger interpretative study in an Afro-American section of inner-city Philadelphia (Ley 1974), and, in the face of injustice and inequality, it was challenging to find a way of writing that

**Fig. 15.1** *Latch-key child in north Philadelphia*

disciplined my own reaction against the oppressions of daily life endured by residents (see Fig. 15.1).

The task of mapping graffiti was not a trivial one, for it raised additional questions – as mapping always should, if it is to pass beyond description and into interpretation. The map should be a tool that poses questions rather than resolving them, the beginning and not the end of scholarship. Cartographic representation is also an act of disclosure and in this study raised its own ethical problems. In order to conceal the identity of individuals, it was also necessary to disguise the neighbourhood, a necessity achieved by maps that included renamed streets and some rearrangement of land uses.

The distribution that emerged from plotting graffiti revealed distinctive spatial patterns coinciding with the disposition of street-gang territories. But it also showed a separate type of graffito that was not so parochially bound but instead sought out highly public sites around the city, including a jet at the international airport and a police cruiser spray-painted while the police officers were inside. These were the markings of what I called graffiti kings, male adolescents who left a highly accented signature of an exotic nom de plume – for example, Cool Earl or Dr Cool No. I. An interpretation of their cultural landscape emerged from participant observation that involved detailed interaction with local youth. In the phantasmagorical world of the graffiti kings there are no social or spatial limits; a displacement from the hard world of street life into a new identity removes the multifaceted exclusion that is an everyday reality, makes all things possible, all spaces available for occupation. To capture with one's signature the most inaccessible site is to make a claim of pre-eminence for oneself, no less real in its own subculture than is residence in a blue-chip neighbourhood in the status world of the elite.

The Philadelphia study raised its own interpretative questions. The gap between researcher and researched was vast, so broad in fact that I became the neighbourhood exotic as local adolescents sought to mimic my British accent when we passed on the street. This gap raised both ethical and methodological issues. What right does the researcher have to intrude into this space of otherness for benefits that are minimal to the marginalized lives that we meet? My own contract involved working as an unpaid community planner for a neighbourhood organization and with a teenage group in the district where I saw my task to broaden the horizons of adolescents enveloped in an environment of limited opportunities. Local residents were employed as interviewers, and interviewees were paid for their time and information.

Equally vexing is the methodological question concerning the validity of an interpretation under such circumstances of cultural dissonance. An important answer is to demonstrate consistency and coherence in an interpretation, but that can still be challenged as the coherence of a distant world-view projecting its own need for order onto the other. I do not reject that objection out of hand, for one's own socialized humanity can never be severed from intellectual work and never be fully known. Access to local culture was achieved through the roles of community planner, youth worker and neighbourhood resident, creating a number of points of entry into local social networks. A second response is to note that participant observers have frequently made considerable use of local key informants to aid the bridging of their own position with local culture, and have occasionally included them as co-authors. My own work depended upon two key informants, a middle-aged woman and an adolescent male, both active in the neighbourhood. Was this partnership enough to bridge the cultural distance that separated my biography from the community? A third and perhaps the most demanding test occurs when the interpreter is held accountable to his or her subjects for the interpretation rendered of them and their neighbourhood. Several years after the study I was invited to make a presentation in Philadelphia, with several community residents in the audience. This is perhaps the most gruelling test of an interpretation, and speaking to the residents later I was relieved to learn of their general endorsement of the story I had told.

## Dealing with diversity: transnational interpretation in Poughkeepsie and Oaxaca

Our second set of examples identifies additional effects of the author's positionality for the crisis of representation.[3] Between 1993 and 1995 I conducted fieldwork with a transnational Mexican community in the small city of Poughkeepsie, New York, and in San Agustin, one of several rural sending villages in the state of Oaxaca. After the first arrival of Oaxacan immigrants in 1984, the migration chain developed rapidly. This movement to and from Poughkeepsie, still characterized by primarily male circulation at the time, was relatively recent yet already a prevalent phenomenon in San Agustin, with most young men and a few women leaving the village around the age of 15.[4] Through ethnographic research I endeavoured to understand the social geography of transnationalism according to migrants' enactment and interpretation of place in their daily lives (Mountz and Wright 1996). After months of participant observation and open interviews, I worked through extensive field notes and interview transcripts to conceptualize complex, collective and contradictory interpretations of change. Everyone, myself included, maintained sometimes contradictory perspectives according to their own experiences of transnationalism. Narratives varied according to personal histories of migration, class, gender, religious affiliation, age and so on. I suggest three strategies to deal with such diversity in interpretation: listen carefully to conflict, maximize polyvocality and contextualize narratives by discussing subjectivity in an effort to situate knowledge. I will illustrate how I used these tools to interpret data that I view as rich *because of* tension among competing narratives.

One strategy of interpretation is to listen for absence, fissure, disruption and contradiction. As Anderson and Jack have advised in listening to women's oral histories, 'An interview that fails to expose the distortions and conspires to mask the facts and feelings that did not fit ... will miss an opportunity to document the experience that lies outside the boundaries of acceptability' (1991: 11). For many, visible changes in San Agustin symbolized a transition to a more successful village where wealth had accumulated via migrant remittances and was displayed in home construction and renovation. Migrants also supported community projects such as road and well construction and sponsored increasingly larger and more costly community festivals to celebrate a variety of events (Fig. 15.2).

Many narratives reflected positively on these changes and rehearsed these symbols of success. A 27-year-old migrant back in the village noted the impact of this visibility:

> 'They see when some guy comes from the United States, you know, he brings money, he makes a nice house, buys the land, everything. So some guys, they go two years, three years, and they make a lot of money. So I think that's why a lot of kids want to go and work.'

**Fig. 15.2** *At the annual festival young women dance to celebrate the patron saint of the village and the return of US migrants to Oaxaca. Remittances have intensified festival traditions, including attire worn and gifts exchanged*

Still others challenged this success and articulated absence in their narratives, noting the loss of loved ones and the financial and emotional stress experienced in absence. Counter-narratives emerged in tension with those who praised the 'model' migrant and pointed to those who had disappeared from the transnational circuit and no longer supported life in the village. Migrants who did not fulfil financial expectations were labelled 'irresponsible', and they and their families were relegated to the bottom of a more complex class stratification that was emerging. Other narratives were motivated by other ideologies, such as a small group of Seventh Day Adventists who articulated disagreement with conspicuous consumption of larger homes and festivals. The reality of migration thus took on distinct meanings in daily life for different individuals, according to the many axes along which subjectivity was formed. Interpretation of change as success in one narrative corresponded with simultaneous articulations of exclusion and impoverishment in another. Ethnographers must be attentive to listen for patterns and their disruptions, for silences in one narrative may be replaced by animated outbursts in others.

Conflicting interpretations of experience within families resonate with this broad tension within the community. Two siblings, Francisco and Marta, occupied neighbouring homes in San Agustin. Francisco and his two sons had worked in Poughkeepsie for several years. Prospering with remittances, his family made considerable home improvements, maintained a well-respected social position, and sent the two youngest daughters to primary school (see Fig. 15.3). Francisco's sister Marta did not experience

**Fig. 15.3** *Large brick homes constructed with migrant remittances from the USA have replaced the much smaller adobe structures that once housed villagers*

transnationalism in the same way. Her husband in Poughkeepsie fell out of contact with the family for long periods of time and rarely sent money to support their four children, who continued to live in a one-room adobe home. She was unable to attend to illness in her family and her children were determined to leave primary school to work. While Marta worked in tortilla production, which continued as a local form of labour in San Agustin, the cost of family support could no longer be defrayed by local income. Rumours held that Marta's husband was an alcoholic who still lived in Poughkeepsie but was unable to keep a job. The narratives of Francisco's family members were shot through with disapproval of Marta's situation and Francisco was always reluctant to support his sister financially. Marta and other women in her situation faced greater challenges because the increasing cost of community reciprocity excluded them, and the disappearance of family members in the USA left them feeling trapped and unable to support their families. Marta's interpretations of change reflected the ways in which her brother's success reduced her hopes for community participation and family survival.

Women such as Marta were often among those who contested the generalized experience of the model migration experience. Anderson and Jack suggest that we 'inquire whose story the interview is asked to tell, who interprets the story, and with what theoretical frameworks' (1991: 11). Many, particularly younger women whose male peers had migrated, felt pressured to stay when they in fact wished to migrate themselves. Consider the following statement made by Maria, a young woman who migrated to New York at the age of 14 and returned a few years later to San Agustin:

> *'Many people, many mothers – in particular the mothers – think that because her daughter is going to go – if she is still unwed, then she has preserved her virginity – they think that upon going there, she can lose it, she can end up pregnant ... Then, when they come back, they are no longer the same. This is the fear of the mothers ... I went, and nothing happened to me. I am complete and healthy. It is the person, not the place.'*

Women's narratives of mobility provided highly complex interpretations of the contested gendering of migration (Alicea 1999). Maria's narrative reveals that she still believes in her role to remain 'healthy' for marriage, yet she challenges other conventional expectations that women must remain in San Agustin in order to preserve this health. She frames her mobility to the USA as successful, challenging masculinist narratives of place as dangerous. Through the process of interpretation, I listened for the ways in which women negotiated external expectations, which relied on masculinist constructions of women as vulnerable and of migration as a male experience across a masculine border into masculine forms of labour and lifestyle in Poughkeepsie. Polyvocality worked as a strategy to present tensions in interpreting the gendering of experience.

Though my own experiences of mobility lay far outside the experiences that I studied, they nonetheless influenced my interpretative process as researcher and author. Many experiences positioned me in particular ways in relation to the data. The city of Ploughkeepsie, where I was brought up, was not the same place to me as it was to Oaxacan residents. For myself and my peers growing up, Poughkeepsie was a place from which many of us hoped to escape. For Oaxacans from rural villages, Poughkeepsie was a place of abundance to which many aspired to arrive. According to one woman who had never been to the city, 'They say that there is a lot of everything there ... a lot of water, a lot of air ... a lot of everything.' How I moved through the village, who I met and interviewed, and how they constructed identity also impacted upon my interpretation of life in the community. Researcher and participants negotiate identities through interaction, a process Gillian Rose refers to as constitutive reflexivity (Rose 1997). My own life experiences shaped my reading of narratives and informed my own identification with competing readings of success. I envisioned success as better health care and opportunities for education, and thus sympathized with those who opposed the quantity of remittances invested in festivals. Though I did not express these views in the field, I did make an effort to represent them when presenting my research.

I was also influenced by the political context in which I conducted this research. It was a time of loud anti-immigration rhetoric in the USA, and simultaneously a period of economic crisis in Mexico with the 1994 devaluation of the peso. Through my research I had developed a personal politics regarding immigration that was at odds with nativist rhetoric, and I responded to that rhetoric through my work. As a woman and a feminist, I sympathized with the women who were marginalized for a variety of reasons by changes that accompanied the transnationalization of the community. This position and politics also informed the tensions I sought to emphasize in my interpretation and presentation of narratives of mobility and immobility. Each of these influences and many others framed my interpretation of data. In order to situate knowledge and interpretation, I worked to contextualize the view by being reflexive in my work about my own position (positionality), by presenting multiple narratives of change and mobility (polyvocality), and by listening carefully for tension among them.

## Conclusion: other storytellers

In this short chapter, our primary task has been to reflect on our own research experiences in addressing the challenges of interpretative fieldwork. We have drawn attention to several issues: the position that all research is interpretative; the authority of our position in the field as privileged problem-solvers with its ethical and epistemological challenges; and the intellectual task of dealing with diversity and resolving competing accounts of the same events.

We have tended to emphasize the social aspects of the crisis of representation in interpretative research. But the researcher is also an individual, albeit an individual bearing the imprint of a larger culture and society, and entering into a research relationship implies personal as well as political associations. In conclusion we refer to several ethnographers whose identification of oppression, grief or injustice motivates a highly charged narrative with explicit personal intervention. Here the humanist belief in a strong author dispels postmodern fears concerning authorship, and the interpreter uses emotion strategically to draw readers into the experience and to relate the individual experience to broader structural issues. Philippe Bourgois (1995, 1996) and Ruth Behar (1993, 1996) are two anthropologists who have adopted this interpretative strategy, Behar from interviews with a Mexican woman who made her living peddling food in a city market, while Bourgois spent his time with drug-dealers in East Harlem. Both Bourgois and Behar insert themselves into their texts in unique and effective ways, which is to say that, by providing autobiographical context, their qualitative research is presented more persuasively. Both invest much of their introductory chapters explaining their positionality in the field and in the text, and how they came to their decisions to shape their interpretations in particular ways. Both are present throughout in the presentation of recorded dialogue and field notes; indeed, Behar's book ends with a 'biography in the shadow', in which she tells her own personal history in the academy.

There is not space here to present in any detail the limitations to interpretative strategies that involve autobiography, reflexivity and emotion, but we should note its risks. Behar, for example, faced adversity from her own family, who objected to her inclusion of personal family matters in publications, and also, poignantly, from the woman around whom she based her entire book (Behar 1995). In addition, there are also instances where these strategies may be overused or inappropriate, and may become 'narcissistic self-reflection' (Bourgois 1995: 14; Behar 1996). So the question becomes where, when and how we draw the important line between reflexivity as rigorous contextualization of qualitative data and narcissistic, emotionally motivated navel gazing. Behar provides important instruction: 'Vulnerability doesn't mean that anything personal goes. The exposure of the self who is also a spectator has to take us somewhere we couldn't otherwise get to. It has to be essential to the argument, not a decorative flourish, not exposure for its own sake' (1996: 14).

We will end with a geographer who has included autobiographical accounts as a strategy in his own ethnographic interpretation, a study that reveals the thin line to be followed between insight and narcissism. In John Western's highly personal ethnography (1992) of Barbadians in London two narratives unfold, the 20-year sojourn of a sample of Caribbean migrants to England, read against Western's own 20-year sojourn as an Englishman in the USA. Most audacious and stimulating are the points of convergence in these narratives, when for example remembering his long walks along the chalk cliffs of east Kent he observes in a moment of flashing introspection: 'What for me is the very essence of childhood home happens to be the essential symbol of English insularity' (Western 1992: 271). There follows in the final chapter, entitled 'Islands and Insularities', a creative movement between his own childhood insularity, the island outlook of the Barbadians, and English/British detachment from contemporary Europe in the light of its self-appointed special status as an island apart.

The chapter ends with Western's astonishing *envoi* to his subjects/friends/fellow island expatriates, the London Barbadians, 'I salute them' (1992: 281), a remark that could be taken as basely patronizing were it not for the linked intersubjectivity he has carefully sought to assemble through the previous pages. And here is the point of his self-projection. Not only is the presence of the author advertised, emotions and all, but so too is the capacity for intersubjectivity, for the sharing of experience in ethnographic encounter. As Bourgois (1996) hears of the brutalizing of a handicapped child inflicted by one of his informants, his mind races in both revulsion and sympathy to the recently diagnosed cerebral palsy afflicting his own son. As Western walks in his mind's eye the Dover cliffs, he reflects on the mindset of *the* islander, the Barbadians' affirmation of '*the* human spirit'. In the solidarities of such intellectual derring-do some researchers see an alternate way of coming to terms with authorial positionality and the crisis of representation in the interpretation of people and place we have discussed in this chapter.

## Notes

1.  The use of the first person here and elsewhere in this essay identifies the role of one of us (in this instance Ley) as researcher. It more accurately ascribes agency and also responsibility to the fieldworker than a passive form of expression, despite the awkwardness of employing the singular case in a co-authored chapter. Later in the chapter we will note when the identity of the field researcher changes.

2.  Readers might wish to compare the interpretation in Ley (1995) with the account of Li (1994),

which covers similar ground but offers a different reading. While I disagree with aspects of Li's interpretation, I will refrain from a detailed exposition here. Nonetheless a comparison of the two accounts provides a nice insight into some of the challenges of interpretation.

3. The first person now relates to Mountz as researcher.

4. The village did have a longer history of migration, beginning with the Bracero programme when Mexican nationals were recruited to work in agriculture in the USA from the 1940s to the 1960s. At the time of my research, more women were migrating from other sending villages than from San Agustin. Since completion of my fieldwork, considerably more women have migrated to Poughkeepsie from San Agustin.

## Acknowledgements

The authors are grateful to Jennifer England and Richard Wright, who provided helpful comments on an earlier version of this essay.

## Key references

Anderson, K. and Jack, D. C. 1991: Learning to listen: interview techniques and analyses. In Gluck, S. and Patai, D. (eds), *Women's words: the feminist practice of oral history*. New York and London: Routledge, 11–26.

Baxter, J. and Eyles, J. 1997: Evaluating qualitative research in social geography: establishing 'rigour' in interview analysis. *Transactions of the Institute of British Geographers* NS 22, 505–25.

Behar, R. 1996: *The vulnerable observer: anthropology that breaks your heart.* Boston: Beacon Press.

Bourgois, P. 1996: Confronting anthropology, education, and inner-city apartheid. *American Anthropologist* 98(2), 249–65.

Western, J. 1992: *A passage to England: Barbadian Londoners speak.* Minneapolis: University of Minnesota Press.

## References

Alicea, M. 1999: 'A chambered nautilus': the contradictory nature of Puerto Rican women's role in the social construction of a transnational community. *Gender & Society* 11(5), 597–626.

Anderson, K. and Jack, D. C. 1991: Learning to listen: interview techniques and analyses. In Gluck, S. and Patai, D. (eds), *Women's words: the feminist practice of oral history*. New York and London: Routledge, 11–26.

Bailey, C., White, C. and Pain, R. 1999: Evaluating qualitative research: dealing with the tension between 'science' and 'creativity'. *Area* 31(2), 169–78.

Baxter, J. and Eyles, J. 1997: Evaluating qualitative research in social geography: establishing 'rigour' in interview analysis. *Transactions of the Institute of British Geographers* NS 22(4), 505–25.

Behar, R. 1993: *Translated woman: crossing the border with Esperanza's story.* Boston: Beacon Press.

—— 1995: Writing in my father's name: a diary of translated woman's first year. In Behar, R. and

Gordon, D. (eds), *Women writing culture*. Berkeley and Los Angeles: University of California Press, 65–82.

—— 1996: *The vulnerable observer: anthropology that breaks your heart*. Boston: Beacon Press.

Bourgois, P. 1995: *In search of respect: selling crack in El Barrio*. Cambridge and New York: Cambridge University Press.

—— 1996: Confronting anthropology, education, and inner-city apartheid. *American Anthropologist* 98(2), 249–65.

Clifford, J. 1986: Introduction: partial truths. In Clifford, J. and Marcus, G. (eds), *Writing culture: the poetics and politics of ethnography*. Berkeley and Los Angeles: University of California Press, 1–26.

Duncan, J. and Ley, D. (eds.) 1993: *Place/culture/representation*. London: Routledge.

Harris, R. C. 1978: The historical mind and the practice of geography. In Ley, D. and Samuels, M. (eds), *Humanistic geography: prospects and problems*. London: Croom Helm, 123–37.

Ley, D. 1974: *The black inner city as frontier outpost: image and behaviour of a Philadelphia neighbourhood*. Monograph Series 7. Washington, DC: Association of American Geographers.

—— 1995: Between Europe and Asia: the case of the missing sequioas. *Ecumene* 2, 185–210.

—— 1999: Myths and meanings of immigration and the metropolis. *Canadian Geographer* 43, 2–19.

—— and Cybriwsky, R. 1974: Urban graffiti as territorial markers. *Annals of the Association of American Geographers* 64, 491–505.

Li, P. 1994: Unneighbourly houses or unwelcome Chinese: the social construction of race in the battle over 'monster homes' in Vancouver, Canada. *International Journal of Comparative Race and Ethnic Studies* 1(1), 14–33.

Mountz, A. and Wright, R. 1996: Daily life in the transnational community of San Agustin, Oaxaca, and Poughkeepsie, New York. *Diaspora* 5(3), 403–28.

Rorty, R. 1991: *Objectivity, relativism and truth*. Cambridge: Cambridge University Press.

Rose, G. 1997: Situating knowledges: positionality, reflexivities and other tactics. *Progress in Human Geography* 21(3), 305–20.

Western, J. 1992: *A passage to England: Barbadian Londoners speak of home*. Minneapolis: University of Minnesota Press.

# PART 6

## Writing

# 16
# Writing conversation
## Katy Bennett and Pam Shurmer-Smith

*This joint chapter is a fictitious recreation of a conversation about writing. It attempts not only to discuss issues of writing but to show the process of writing and how events, emotions, histories and moments affect it (see Fig. 16.1).*

**Fig. 16.1** *Katy Bennett and Pam Shurmer-Smith, Old Delhi railway station: time out from ethnographic fieldwork*

## Out of order

### Earth

Dear Pam,

I didn't feel that our recent get together was quite the same. Mother-in-law was on her way. No time to relax with you over lunch. No time to talk. To really talk outside of your book, my research, your job, my job. John amazes me every time I see him – he's so tall and grown-up. He'd hate me for writing this, but I still remember him on the field trip when I was at the back of the bus and you, Louis and he were at the front. He was little and made us all laugh whilst we stood against a marshy Breton landscape listening to French voices and then English voices talking to us from the front, suddenly broken by one voice from the back saying 'Mummy, I'm beginning to lose my patience now'.

I can't remember the last time I wrote to you. We used to write to each other so often whilst we were away on fieldwork – me in Dorset, you in India, me worrying about not observing, writing or constructing an ethnography properly, you answering, reassuring, advising. Now it's the occasional postcard or email – not the same as those letters from India, which travelled all that distance to reach me in Melcombe Bingham where I was writing about farmers' wives, patriarchy, kinship, consumption ... for my Ph.D. thesis. We were in such different places. I was working and living with a farming family, you were working and living in Lal Bahadur Shastri National Academy of Administration in Mussoorie. I felt a stranger, you at home – which was silly, because it should have been the other way round.

Sorry for not writing. No excuses. Apart from being caught up in trying to ground our project and write it into the confines of a report. We're allocated so few words which we're trying to fill with so many voices. I've probably already told you this, but we interviewed over 70 key informants representing the formal sector, all of whom were involved in the design and delivery of coalfield regeneration strategies. Layered onto such voices were those of the community sector whom I visited, watched, listened to and sometimes interviewed whilst living with ex-mining families who themselves told me more stories. All these different, sometimes conflicting, occasionally contradictory stories wanting to be told and yet we're confined to writing a sensible, smooth text. A text made up of short, simple sentences, which the reader can speedily grasp and understand and use to tell another story. Their story. We're allowed no contradictions, no messiness, no life – just dead easy text. 'Dear Katy', she wrote, 'You may need to look at the structure of the report in relation to the length – and focus on the key messages that you want to get over. I think you will also need to be careful about language. "Problematizing" and "visibilize" may be OK for sociologists but not for the rest of us!!' Not your voice, our voice (and how I hate those exclamation marks). I'm learning to write differently Pam. I'm having to tell and not evoke and as I tell, I'm moving further away from the clamouring voices of the researched. I can hardly hear them now as I decide which are the key messages and how to set them out in simple paragraphs. I don't feel grown up enough to tell.

Ideas are beginning to take shape in my head for our chapter for Claire and Melanie – maybe we can start writing to each other again in preparation for our 'writing conversation'. Maybe we can somehow show the process of writing. I'm trying to ground my ideas and see how they look in writing but 'words have a habit of pinning meanings down, making the complex simple and the uncertain fixed ...' (Valentine 1998: 308). As you know, I find it difficult to show what I mean through words and language.

How's THE book coming along? It sounds as though it's so nearly there – just that one difficult chapter to write. You'll get cross with me for writing this, but I know it'll be good.

Louis knows too. I hope he doesn't feel too harassed by me at the moment. It's just that the girls are so keen to plant a tree for Phil outside the geography department – they keep asking Giles and me if it's possible.

I must go – I have a meeting.

Love Katy.

### Fire

Dear Pam,

I read this and liked it. It's from Helene Cixous's *Stigmata*:

> *I was thinking about the challenge that life and death lay down to writing. For writing moves at the pace of the hand. And life and death go by in a flash. We catch fire, surprised. Writing is far behind. How are the fiery moments to be grasped? How can that fire be caught in our hands? Besides, we had better catch the fire in our hands quickly. Because if we catch the fire in our hands quickly enough, we won't get burned. Fire grasped with the speed of fire: that's how it must be done. To write. (1998: 39–40)*

My days of talking to you about Cixous and Irigaray seem a million miles away now. My frustration, struggles and fears of not understanding. I've started reading Cixous in the evenings before I go to bed; my days are filled with regulation theory, turf wars and governance. I still find her difficult, sometimes she escapes me, sometimes she makes blissful sense – writing about things and feelings that I thought were impossible to write. I know she wants to escape me, she would hate to be entirely captured. She's fiery.

For me Cixous shows the failures of writing ... how can I ever capture the speediness of life, the voices of the researched in my text? Writing can never tell how it was; it is a dynamic creative process, it provides a way of discovering at the same time as creating new meanings. New meanings that draw on the feelings, relationships and thoughts of the writer as she tries to recount the stories of the researched. I can use Laurel Richardson to back me up on this: 'Nothing is simply present or absent but ideas are in transformation; "facts" are interpretations "after the fact". Self knowledge is reflexive knowledge' (1993: 704).

I don't think representation is ever a possibility. I might have tried to represent – but I never succeeded. And who wants to represent anyway? It involves being (doing?) powerful and capturing the words/lives of the researched. Impossible. I'll always fail ... even when I'm meant to be writing powerful reports I don't represent, but re-present. The hyphen provides a pause during which meanings shift and stories evolve. Re-present is less smooth than represent; it physically, clumsily, shows that there has been a shift in meaning, that what is read wasn't what they (the researched) said. It doesn't capture and contain them.

Far better than to re-present, to try to represent, is to evoke. (I'm thinking about Tyler's chapter (1986) now.) To evoke is sensuously to depict, requiring the writer to move beyond re-presenting what is visual; requiring the writer to move beyond describing what was/is looked at and objectified (Clifford 1986). Evoking requires the writer to have subjectively engaged in a more sensuous experience and to write down, up, out (from her entangled positionings) what she smelt, felt, heard, tasted ... sensed ... At best I evoke, at worst I end up re-presenting.

Pam – I was going to save this for your birthday card, but I'll write it now. All this thinking about writing is reminding me of you. Reminding me of my luxurious Ph.D. days (the blood, sweat and tears have slipped from my memory), when I had time to think about writing, how to write and how to construct an ethnography. How to 'do' an ethnography. Remember those letters ... 'Katy – you must look for age ...' So here's what Cixous has to say about age:

'Our age', 'if it counts',

> is for our mother, it's our mother who is it, it's she who keeps count. We, we are always interiorly our secret age, our strong-age, our preferred age, we are five years old, ten years old, the age when we were for the first time the historians or the authors of our own lives, when we left a trace, when we were for the first time marked, struck, imprinted, we bled and signed, memory started, when we manifested ourselves as chief or queen of our own state, when we took up our own power, or else we are twenty years old or thirty-five, and on the point of surprising the universe.
>
> The other age, the one which our mother authenticates, remains forever foreign to us, and yet, bound to the mother, it is sacred to us. (1998: 64)

Love Katy.

## Water

Dear Pam,

Maybe through writing our chapter for Melanie and Claire we can conclude our video. Not literally, I know. It's already finished. Remember the hilarity of filming our (dead) bodies for its opening shots? 'The Body in the Field' (Shurmer-Smith and Bennett 1995). That was the start. This isn't the end, but maybe the conclusion of that part of us that we wanted to see through to the end.

I've been thinking about how to write evocatively, how to show an audience or a reader an understanding or an interpretation of the research (situation). It's easy to problematize (problematize, problematize, problematize), to talk about evocative writing but less easy to do. It takes guts and a risk of failure and being laughed at. I've been rereading some of the letters you sent to me whilst I was doing my fieldwork. I was in Dorset and you were in India and clumps of letters would arrive at a time – some would take a few days, others weeks – so they'd arrive in clumps and we'd be replying and asking all out of sync. Messy, entangled writings – talking at tangents. Making sense. Contradictions. A bit like now really. You're going to get these letters in a clump too. Just like when you were in India.

Remember writing this (25 April 1995 – Mussoorie)?:

> *Your experiences seem absolutely typical, masses of stuff crowding in and difficulty knowing how to write it all up (followed by asking yourself why you thought it worth writing down in the first place, half the time!). It's impossible in the field to know what's important, as you never know what will lead to something interesting and what just fizzles out; which people turn out to be crucial in a connection between say two cliques, even though they don't seem terribly important in themselves.*

Rereading your responses to my problematizing takes me back to the angst of not knowing what to write down, what ought to be written, what's the most important part of a conversation, an observation, an event. Important is not really the right word. The key that unlocks. It's easy for me in retrospect to glide over this part of writing conversation, to forget the feelings of drowning in voices and 'data' and not knowing where to start. Keep writing you said. Just keep writing.

Writing is about creating meaning.

Through my selves. Through language, which isn't innocent either.

I am getting so annoyed. All I'm doing now is writing and deleting, writing and deleting, writing and deleting. I'm getting cross with myself because I'm unable to write what I mean. Half my problem is that I keep wanting to write 'Duncan and Ley (1993) showed that blah, blah, blah', but I don't want to revert to attempting to 'do' the all-knowing academic. I don't want our writing conversation to be a standard text that does the usual of setting out our writing plan against a heritage of what's been previously written on the subject. But Duncan and Ley (I can't help myself), in their introductory chapter, point out to me that there is nothing new about what I'm attempting to do here. They are critical of what they term 'postmodern' experimental and polyphonous writing and much prefer an interpretative writing based on hermeneutics, which acknowledges the role of the interpreter in writing and also spits in the face of mimesis.

Interpretative writing is beginning to take up space in geography texts as writers recognize the partiality of their viewpoints, which are clouded by the purpose of their viewing in the first place, clouded by their judgements, background and positions in life.

I'm drawn, though, to a way of writing that not only recognizes partiality and the role of the author in the construction of a text, but that has a better attempt at evoking meanings and understandings. Writing that tries to show, not tell, and where the meaningful process is as important as the final product. Fluid writing. Sometimes I might not be understood, and this is certainly not my intention, but occasionally I might make more sense. I like swimming against the tide. I don't know many geographers that have attempted experimental writing. In geography, Olsson and Pred are normally recognized and captured in references.

I've found myself turning to other disciplines.

Laurel Richardson (1994) helps me by usefully mapping out the different (sociological and anthropological) attempts at trying to write evocatively, attempts at writing unconventionally so that the reader might also glimpse what the researcher felt, saw, heard ... It's a good place to start. She considers different genres of writing which she describes as 'experimental representations' (1994: 520). She begins with 'the narrative of the self', a form of writing that is deeply personal and attempts to make sense of issues through the author's own lived experiences. The author attempts to carry the reader along with her, so that the reader might experience for herself some of the issues that concern the writer. It's a sort of writing that opens itself up and invites the reader to step inside the text, to feel for herself. Like in a film, when the narrator carries me, the audience, along with her, so that I too can begin to feel what she experienced ... Maybe this is a squirmy sort of writing, but sometimes it can succeed. Carolyn Ellis (1993) movingly carried me along with her when she recounted the event of her brother's sudden death caused by a plane crash (when he was on his way to visit her). Her writing raised all sorts of issues about death, life, families, mourning and relationships that might not be easily captured in an attempt to write more objectively through the issues. Maybe I found Ellis's paper so engaging right now because it relates to my feelings about Phil's death. He wasn't my brother, but we were close. I know you didn't like him when you met him at my flat – he was capable of making me hate him too. I loved him though. I now only remember the best times, the good memories; bad ones have slipped away, which makes coping with his sudden death that bit harder.

Ellis and Bochner (1996) call this auto-ethnography. (I had quickly to change the subject away from Phil.) A whole section in their book – 'Composing Ethnography' – is labelled as this, where contributors overtly weave their text through their selves when talking through issues like their bulimia, retarded mother, breast cancer and health-care rationing. In all of the papers meaningful issues are raised and evoked as I swim back and forth feeling different from and similar to the author. Occasionally it all seems too dramatic. But writing is dramatic because it is about making a point, selecting, labelling and cementing through language.

Richardson (1994) moves on to talk about ethnographic fictional representations, poetic representation, ethnographic drama and ends with writing about mixed genres (like Margery Wolf's *Thrice-Told Tale* (1992), which is written in different ways, with each way throwing up different meanings). There are lots of examples of these. Richardson (1992) turns her transcript from an interview with an 'unwed mother' called Louisa May into a poem, which evokes the way Louisa May talks, the conversation itself and the issues it raised as well as Louisa May's philosophy for life – 'so that's the way that worked out'. Mienczakowski (1996) shows the meaningful production of 'Busting' and 'Syncing

Out Loud', two ethnographically based plays exploring the experiences of persons undergoing detoxication processes and experiences of schizophrenic illness respectively. Each production is layered with all the (different) understandings of the researched, script writers, actors and audience. Fox (1996) explores issues of child abuse through a three-person account, with her own voice at the centre (in both senses of the word) and that of the sex offender and victim either side of her.

Is this all shaping up in the right way?

Love Katy.

PS You're the only person with whom I end my conversations so abruptly. Most people say 'take care', 'love to Giles', 'look after yourself'. But you're different. We can be chatting on the phone for ages, or writing long letters or e-mails – but when it's time to end them, it's always suddenly. I've never quite worked out why. I feel that I have to do the same.

### Wind (Air)

Dear Pam,

I've just been chatting to a friend about the use of video in research. We ended up talking about how video ends up showing more about 'me' than it does about 'them' out there. Maybe video does more overtly what writing manages to hide.

Actually, I've had quite a few conversations with people that have affected my thoughts on writing. After a seminar yesterday I was talking to someone about writing conversation, writing up interviews and public hearings. We were thinking that writing up a public hearing might be best done through a drama or play – a form of representation that might better evoke the different voices, opinions, ideas and the conflicts. I talked about my attempts to write each chapter of my thesis differently, in a way that would better evoke what it was that I was attempting to write. I explained that one chapter was a story – a fiction in the way I collapsed time and shifted from one person to the next as each person's view of fiction was articulated. The details of the story were true in the sense that I stayed as close as possible to how I remembered them, wrote about them in my diary, or transcribed them. And then I came round to the play – which attempted to show an interview context, intersubjectivities, shifting friendly relations with feelings of space widening and closing according to a particular moment of the interview, according to how the interviewee is engaging with the interviewer and vice versa.

My biggest fear with writing conversation is that novelists, poets and 'proper' writers can do all of this far better than me. They often have an ability to use words to evoke what is impossible to evoke, to 'capture' the impossible. To shape with their words that which previously had no form. But I like the idea of sometimes trying to escape academic structures to show things differently, to get away from the academic path of knowing to explore alternative routes (roots?). Who knows then what might be 'seen'? And who knows how this might help/contradict/mess up academics' (way of) knowing?

Some of this might be a bit too much for our chapter Pam. But hopefully what we do eventually write might open up the act of writing conversation, problematize it and show its possibilities. Mike Crang (1998) asks: 'if anyone were to look around for accounts that really gave the reader a feel for a place would they look to geography textbooks or novels?' He tells us that 'the answer does not need saying. Undergraduate geographers receive years of training which seem to remove the ability to write a piece of prose (let alone, say, poetry) that imaginatively engages its reader' (1998: 45). I don't know why geography as a discipline is like this. Perceived expectations. Fear. Valentine (1998), when it seemed she had nothing more to fear, rocked when she wrote her paper that began to open up all sorts of issues through her 'hateful' experiences. Poetry, fiction and drama only really enter into geography texts when invited – guest text. A poem appears in *Society and Space*'s editorial (1998). It is called 'raise shit' and is written by Bud Osborn, a poet and activist from Vancouver's Downtown Eastside. The poem was presented in a workshop – 'New York–Vancouver, Gentrification and Memory' at the Inaugural Conference in Critical Geography held in Vancouver (August 1997). 'raise shit', whilst it evokes the different (perceived) faces of Downtown Eastside, is critical of gentrification whereby:

> the wealthy move the boundaries
> and the poor have to keep out of the way . . .

Nick Blomley, in his introduction to the poem, states that 'one participant in the session described it as the best "paper" he heard at the conference, which is both a compliment to Bud, and an implied criticism of the work of many self-defined academics' (1998: 279).

Love Katy.


Dear Katy,

Yes, I remember getting your letters out of order and giving up on the possibility of running a conversation. It was a bit unreal giving supervisory advice when I was trying to work out how to get past the polite disinterest that surrounded me. I don't think you realized that one of the reasons I went off to do fieldwork was that I was conscious that, though I was advising you all about participant observation, it was nearly 20 years since I had submitted to its rigours. As to your being lost and my being at home, it is interesting to contemplate why you needed to believe that. I have since revealed (Shurmer-Smith 1998) that I was very much an outsider in Mussoorie until the end of that first stint of fieldwork, whilst you were on your return visit to Dorset. But, yes, there was a sense in which I was comfortable with getting on with being a stranger.

When I did my own Ph.D. research I did not have a close relationship with my supervisor, much as I admired him – he took a sabbatical to help run the Portuguese revolution and I thought my concerns were trivial by comparison. Eventually he left LSE to become a diplomat and a series of step-supervisors took over. I'm not sure that I regret the lack of advice. Anthony Cohen drew a gasp from the audience at the 'Anthropology at Home' Conference of the Association of Social Anthropologists when, as rapporteur on Nigel Rapport's paper (1986), he said, 'Sometimes I feel like a midwife who looks down at the product of her work and wonders what awful monster she has helped to bring into the world.' Both former supervisor and supervisee glowed with pride and affection!

What 'awful monster' takes three and a half thousand words of a shared five-thousand word allowance, dictates the structure and then has the cheek to talk about problems of the clamorous voices silenced by other people's representations? No, I will not demand my 'right' to represent myself, though I don't easily recognize the 'me' you represent.

Certainly representation is always a problem and, unless one is very arrogant, writing about other people makes one nervous. Actually, writing about anything is pretty nerve-racking, since it exposes not only one's ignorance, but also one's peculiar knowledge. Perhaps unfashionably, I think that all representations are true (to something or other) and that readers fairly regularly plug into those that are truest for them. I don't believe that the people being represented necessarily have a superior claim, though often they need to fight their corner. Don't you think that the distinction between 'representing' and 're-presenting' is a bit of a sophistry? I understand the discomfort that prompts you to refer to it, but that is another matter.

That word 'represent' is also used in the context of standing in for. A defence lawyer is not expected to reveal what really happened; she is supposed to draw upon specialist knowledge to put the best possible interpretation on things and will give explanations the accused would not generate. This representing implies taking responsibility for what one says for someone else and comes quite close to what I believe a piece of academic writing based on qualitative research methods should aim at. Representation is not just presenting again what was already presented; it is about coming up with new thinking about human relationships. As far as I am concerned, if there is any point at all in the academic enterprise, it is this new thing revealed through comparative study, training in theory and sensitivity to one's own powers of observation. I don't think that the writing task is primarily to tell readers what it is like for a few hundred people in a small place.

The excitement of fieldwork comes from talking to the people I'm studying (with) and seeing whether my views are greeted with enthusiasm or dismay. When it is enthusiasm, it is unlikely to be because I have imitated their own thinking but because I have offered an interpretation that makes sense to them. Then, as in a good seminar, we can go on together to build more interpretations, genuinely constructed out of the interaction. When people register dismay, I become pretty dejected, but listen to why they think I am wrong; this always generates a new perspective (but sometimes it actually reinforces adherence to my own interpretation). Much of my fear of publishing, and the preceding writer's block, comes from the realization that this version will stand, even if corrected later, and that I have claimed the last word in the dialogue. This does not just apply when writing ethnography; when I was writing the book you refer to, I was always conscious of a small number of people in India who had helped me to formulate my viewpoint. I did not want to let them down.

You have heard me before on the tyranny of the dictaphone (best exemplified in Visveswaran 1994), which leads uninspired researchers to concentrate on capturing as many words as possible, often totally out of any context other than that generated by the interview. Research for research's sake. Mass production research. I get angry inside when people ask me how to write this stuff up. What they mean is how much and which bits should they appropriate and how can they pad it out to claim it is 'academic'. The dictaphone has its uses, but I believe that the writing has to be the author's view and the author has to have gone through the process of surrendering to the ethnographic situation. Ethnographic research should transform the researcher and it certainly is not for people who are unwilling to take risks with their selves. I also believe that one writes what one becomes and simultaneously clarifies the process of becoming by writing. This is not just writing up, but writing all the time – writing down one's thoughts and feelings, writing out to other people. That is what ethnography is, writing people. That is also why Cixous is so important to me. Not what she writes but the way she writes, the fact that she sees writing as so much more than representation, sees it as a way of discovering and a form of creation.

Evocation! I love the idea of calling forth new understandings or angry rejections that expose contradictions; these evocations are like summoning up dangerous forces. I am less happy with the softer sense of evocation, which I see as mouldy with nostalgia. Quite frankly, I don't think it matters whether the vehicle is a poem or a play or a formal report. If what one is trying to say is exciting, it will shine through any medium (even tables of statistics). If it is not exciting, poetry will only make it more embarrassingly banal. A poem from someone involved – great; a poem from someone who wants to look involved – cringeworthy. New modes of representation from mere amateurs – forget it! The best you will get in response is politeness. Cixous knows what she is doing, she has done her apprenticeship in multiple languages and in the theatre and she knows how to fall without breaking her neck.

A few years ago a journal doctored a book review of mine, replacing 'I think' with 'In this reviewer's opinion'. I was livid. I don't write in this scientist third person style and I fall about laughing when people write about their fieldwork as if someone else with the same name happened to do it (whilst they watched from a cloud). The way we write ourselves is important, but that is not the same as saying that ourselves are important. Usually in academic writing, our own personal relationships and emotions are just 'noise' as far as readers are concerned. Readers are not a captive audience for our shaky sense of self and we should introduce this only when it is part of the argument. I feel uneasy at finding my family and your friends in this chapter, because I do not feel they are part of the issue of representation, which is our brief. It is difficult to write this, because it will hurt you, but a friend's death is rarely relevant to the act of representing research. For me it feels dragged in, jarring, even unfair and exclusionary of your readers. There is always a tiny part of our writing that is not for our readers at all, but just for ourselves, but this needs to be managed very carefully so that it remains in the private realm in which it was generated. (This can be done by using words that someone special used or inserting a reference that was part of a shared argument).

It is now acknowledged that all ethnography is fiction in the sense that it is the product of the author (Strathern 1987). The least we can do as authors is shoulder the responsibility and tell it the way we see it. If we can't see 'it' clearly enough, perhaps we should just stay silent.

You did say I always end abruptly,

Love,

Pam

PS I am 27 words over my 1500 – shake them out of your bit!

## Postscript

*From Katy: Before I wrote this, I had a picture in my mind of how I hoped the words might map out, how I hoped they might show our different thoughts and ideas on writing. I love words because they can be so meaningful and I can try to share those meanings with people I find irresistible. Writing this turned into a painful experience, because it captured certain feelings, showing me what I didn't know was there and clouding up what I thought made sense. So much has been written out here. And so little. Occasionally words don't make any sense at all.*

*From Pam: The pain Katy refers to, for me at least, emerged from having to confront the realization that, as we scratched away at the meaning of research and the way in which it could be communicated, we were starting to uncover the probability that, though we had been agreeing for years, we had quite radically different positions. We both value subjectivity in research, but we come to the self (our own and those of other people) from radically different perspectives. This realization throws into question our right ever to tell the stories of others; Margulies's recent play* Selected Stories *probably illustrates this better than we can.*

## References

Blomley, N. 1998: The poetic geography of gentrification. *Environment and Planning D: Society and Space* 16(3), 279.

Cixous, H. 1998: *Stigmata: Escaping Texts*. London: Routledge.

Clifford, J. 1983: On ethnographic authority. *Representations* 1(2), 118–45.

—— 1986: Introduction: partial truths. In Clifford, J. and Marcus, G. (eds), *Writing culture: the poetics and politics of ethnography*. Berkeley and Los Angeles: University of California Press, 1–26.

Crang, M. 1998: *Cultural geography*. London: Routledge.

Duncan, J. and Ley, D. 1993: Introduction: representing the place of culture. In Duncan, J. and Ley, D. (eds), *Place/culture/representation*. London: Routledge, 1–21.

Ellis, C. 1993: 'There are survivors': telling a story of sudden death. *Sociological Quarterly* 34(4), 711–30.

—— and Bochner, A. (eds) 1996: *Composing ethnography: alternative forms of qualitative writing.* London: Sage.

—— and Flaherty, M. 1992: *Investigating subjectivity: research on lived experience.* London: Sage.

Fox, K. 1996: Silent voices: a subversive reading of child sexual abuse. In Ellis, C. and Bochner, A. (eds), *Composing ethnography: alternative forms of qualitative writing.* London: Sage, 330–56.

Hastrup, K. 1992: Out of anthropology: the anthropologist as an object of dramatic representation. *Cultural Anthropology* 7, 237–45.

Kolker, A. 1996: Thrown overboard: the human costs of health care rationing. In Ellis, C. and Bochner, A. (eds), *Composing ethnography: alternative forms of qualitative writing.* London: Sage, 132–59.

Margulies, D. 1998: *Collected stories: a play.* New York: Theatre Communications Group.

Mienczakowski, J. 1996: An ethnographic act: the construction of consensual theatre. In Ellis, C. and Bochner, A. (eds), *Composing ethnography: alternative forms of qualitative writing.* London: Sage, 244–64.

Olsson, G. 1992: Lines of power. In Barnes, T. and Duncan, J. S. (eds), *Writing worlds: discourse, text and metaphor in the representation of landscape.* London: Routledge, 86–96.

Osborn, B. 1998: 'raise shit'. *Environment and Planning: Society and Space* 16(3), 280–8.

Pred, A. 1990: In other wor(l)ds: fragmented and integrated observations on gendered languages, gendered spaces and local transformation. *Antipode* 22(1), 33–52.

Rabinow, P. 1996: Representations are social facts: modernity and postmodernity in anthropology. In Rabinow, P., *Essays on the Anthropology of Reason.* Princeton: Princeton University Press, 28–58.

Rapport, N. 1986: Cedar High Farm: ambiguous symbolic boundary. An essay in anthropological intuition. In Cohen, A. (ed.), *Symbolising boundaries: identity and diversity in British Cultures.* Manchester: Manchester University Press, 40–9.

Richardson, L. 1992: The consequences of poetic representation: writing the other, rewriting the self. In Ellis, C. and Flaherty, M. (eds), *Investigating subjectivity: research on lived experience.* London: Sage, 125–40.

—— 1993: The case of the skipped line: poetics, dramatics and transgressive validity. *Sociological Quarterly* 34, 695–710.

—— 1994: Writing: a method of inquiry. In Denzin, N. K. and Lincoln, Y. S. (eds), *Handbook of qualitative research.* London: Sage, 516–29.

Ronai, C. 1996: My mother is mentally retarded. In Ellis, C. and Bochner, A. (eds), *Composing ethnography: alternative forms of qualitative writing.* London: Sage.

Shurmer-Smith, P. 1998: Becoming a memsahib: working with the Indian Administrative Service. *Environment and Planning A* 30(12), 2163–79.

—— 2000: *India: globalization and change.* London: Arnold.

—— and Bennett, K. (1995) 'The Body in the Field' at the Annual Institute of British Geographers Conference, University of Northumbria, January 1995.

Strathern, M. 1987: Out of context: the persuasive fictions of anthropology. *Current Anthropology* 28, 1–24.

Tillmann-Healy, L. 1996: A secret life in a culture of thinness: reflections on body, food and bulimia. In Ellis, C. and Bochner, A. (eds), *Composing ethnography: alternative forms of qualitative writing.* London: Sage, 76–108.

Tyler, S. 1986: Post-modern ethnography: from document of the occult to occult document. In Clifford, J. and Marcus, G. (eds), *Writing culture: the poetics and politics of ethnography.* Berkeley and Los Angeles: University of California Press, 122–40.

Valentine, G. 1998: 'Sticks and stones may break my bones': a personal geography of harassment. *Antipode* 30(4), 305–32.

Visweswaran, K. 1994: Feminist anthropology as failure. In Visweswaran, K., *Fictions of feminist anthropology.* Minneapolis: University of Minnesota Press, 95–113.

Weil, S. 1989: Anthropology becomes home: home becomes anthropology. In Jackson, A. (ed.), *Anthropology at home.* ASA Monographs 25. London: Tavistock, 196–212.

Wolf, M. 1992: *A thrice-told tale: feminism, postmodernism and ethnographic responsibility.* Stanford, CA: Stanford University Press.

## 17
# From where I write: the place of positionality in qualitative writing
## Ruth Butler

### Introduction

*'I was born with a visual impairment.'* I have often wondered why I chose to open my Ph.D. thesis with these words. When working on visually impaired people's experiences, there seemed to be a need to make clear my own impairment, but why that was so has sometimes escaped me. What use is such information to the reader? Working within a feminist, emancipatory research framework, using qualitative research methods, it seemed important to state my positionality relative to my interviewees. It was necessary as a means of making clear my awareness of the power relations operating upon myself and the research participants during the design, data-collection and analysis stages of the research process. It has, however, occurred to me that there are some shortfalls in relation to these issues in my writings.

The people I have worked with have come from a wide range of social, economic and political backgrounds, but I have not in any publication stated my age, religion, class, sexuality or numerous other personal characteristics. Indeed, despite reflecting upon the heterogeneity of visual impairments, I have never in any detail explained the extent of my own impairment, its medical cause or long-term prognosis, amongst other factors. Does this lack of autobiography in my work matter? Is my impairment the most important element of my persona to note when working with other visually impaired people? Are power relations between researchers and those they research too complex, working on too many different levels, to cover in any detail even in a lengthy thesis? Is positionality merely given lip service in academic writing? Are such references to the author of value to the reader? Should authors have a more prominent position in their texts or should they rather be invisible?

In this chapter I do not mean to offer any definitive answer to these questions, all of which are worthy of much lengthier debate than is possible here. Rather, through an exploration of some of the issues involved, I look at a range of choices with regard to written style you as an author can make. In particular, I draw attention to the need to reconcile different styles with the varying ethical and philosophical positions you may take in your qualitative research projects. I hope this will provoke further questions and cause you to reconsider and address your positionality as appropriate in your writing and your broader research practices.

The chapter falls into four main sections. First, I outline the 'traditional', scientific structure or formal approach to writing an academic article. I then briefly discuss feminist developments in the philosophy of geography, which embrace positionality and challenge the formal approach to writing. Next I consider two aspects of this question in relation to writing. First, I look at how the accounts given by those who are researched can be written into dissertations and theses. I then discuss how you might include the presence of you, the researcher, in the written product of your research. Finally, the chapter concludes by questioning how theory, ethics and writing styles may be reconciled – offering a number of different issues for further thought.

## Formal approaches to writing

There is a proliferation of texts addressing the issue of how to write an academic document (see e.g. Phillips and Pugh 1987; Gilbert 1993; Greenfield 1996a). They detail at some length the need for common characteristics such as structure, strong supportive referencing, the justification of points and the correct use of the English language. In the context of grammatical structure, the use of the personal pronoun is given particular attention. Whilst acknowledging that different styles are necessary for different audiences – more informal for magazine articles, for example – Levy (1996: 254) suggests that 'research reports should tend towards the formal'. The responsibilities of investigating and sharing with others information about the social, economic and political worlds in which we live are not to be taken lightly. For this reason 'the danger of misinterpretation needs to be minimised. A clear, logical style is therefore appropriate' (ibid). Levy (1996: 254–5) goes on to argue that:

> the formal style can aid understanding ... because you are writing in a scientific tradition and the formal style has been in place in that tradition for over two centuries. As a result it is embedded in the scientific community. It also has the advantage of ensuring
>
> * logic
> * structure
> * clarity
> * precision.

In this vogue distance and objectivity are encouraged in the written presentation of findings (Gilbert 1993; Greenfield 1996b; Levy 1996). The purpose of these instructions is cited as being to clarify your argument and aid its communication to your readership; to give 'clear and effective presentation' of your research, so that it can be 'properly understood' (Levy 1996: 253). Dispassionate, and supposedly unbiased, analytical interpretation of the research lies at the centre of such publications. One way in which this is often reflected is through the formal use of the third person rather than the personal pronoun. Hence, 'the researcher conducted in-depth, unstructured interviews' is seen as preferable to 'I used in-depth, unstructured interviews'. In this way the voice of the researcher has, by design, become distant in the research that an article reports. Such writings have equally allowed little space for the voices of those researched. The predominance of quantitative, distant, research methods in the past, meant that numerical summaries have often subsumed individuals' own words as recorded in more qualitative, in-depth interview settings. The cost of such practices, as Oliver (1994: 64) notes in relation to research with disabled people, is that 'sterile', 'head-counting' procedures have done little to explain the social practices that result in the recorded statistics.

Although formal writing styles like those that Levy promotes continue to have influence on the form of research articles today, it is misleading to say that some changes in practice have not been widely embraced. The vast majority of notes for contributors in academic journals now include the insistence on politically correct language. For example, *Area* states in its notes for authors: 'Papers should be written in non-sexist, non-racist language (see guidelines in *Area* 23(4), 290–94).' With the increased use of qualitative methodologies it has equally become common practice in the social sciences to incorporate direct quotes from interview transcripts in publications (see e.g. Butler and Bowlby 1997; Skelton and Valentine 1998: chs. 3, 4, 5, 12). These changes mark a questioning within geography, and other social sciences, of the philosophical basis of knowledge and knowing. In particular, feminist theorists have criticized the dominant epistemological basis of knowledge and raised the question of social positioning.

## Feminism and positionality

My own reflections about positionality are influenced by writings within feminist theory. Post-colonial feminists, and others, have recognized that all knowledge is produced within certain economic, political and social circumstances, which inevitably shape it in some way (Rose 1997). They have critiqued claims that neutral, universal knowledge has been presented in past academic texts. Most notably, early feminist theorists strongly critiqued what they saw as the male-biased and hence inaccurate theories of the world. However, work promoting a singular, collective women's view has equally come to be seen as unreflective of the world we study. White, middle-class, educated feminists have been accused of creating a partial, situated knowledge, unreflective of the range of experiences of women from different classes, ethnicities, sexual orientations and so on (e.g. Carby 1982; Amos and Parmar 1984).

Work on marginalized groups has drawn attention to differences within populations that have formerly been considered to have a single unifying identity. The misleading and unhelpful nature of dichotomies such as those of male/female, black/white, homosexual/heterosexual and disabled/able-bodied have been problematized by an awareness of the numerous social, economic and political axes – class, gender, race, age, sexuality and (dis)ability, amongst others – that cut across such simplistic binary divisions and are themselves fluid concepts (Butler 2001).

In the light of this literature, feminists have 'recognized the need to situate their discourses within multiple and often contradictory fields of power' (Andermahr et al. 1997: 218–19). Researchers have become increasingly aware of the complex effects of their multiple social, economic and political positionings relative to their research participants as epistemological acts. The need for us all to be aware of the situated and limited nature of our work in piecing together our understanding of society has been pointed out by several individuals (see e.g. Haraway 1988; McDowell 1992; Rose 1997).

Earlier chapters in this book have illustrated the impact of such philosophies on field methods and data analysis, but how do they impact upon writing? How can you show an awareness of the impact of your various positions relative to your research subjects in your writings? Is this, as Rose (1997) suggests, a difficult, or even impossible, thing to do?

## Considerations of positionality in writing

Developments in qualitative research methods have attempted to address the apparent shortfalls in distant, scientific research processes in the field through the increased involvement of the researched at all stages of the research project, as discussed earlier in this book. Emancipatory, collaborative and unexploitative research has generally been seen as synonymous with qualitative methods (McDowell 1992; Stone and Priestley 1996). In a collaborative research project it is argued that the researcher and researched can come to an understanding of what is taking place around them and develop a sense of trust to share their experiences in an atmosphere of safety and support. Commonalities between the researcher and researched can be recognized and become part of a mutual exchange of views (McDowell 1992).

However, these practices do not always come across in the written findings of research. Word limits on reports often reduce methods sections to a minimum. For example, in our own 'Bodies and Spaces' Sophie Bowlby and myself offer only a brief paragraph (in a footnote) to explain the source of the qualitative data cited (Butler and Bowlby 1997: 421). This requirement to abbreviate the full detail of methodologies means that ethical conduct and motives for research practices, including writing style, often have to be taken for granted. In this section I want to reflect on some of the different ways in which researchers have tried to make the subjective voices of research subjects and researchers visible in texts in order to clarify whose views are being put forward in a report and from what position.

### The voices of the researched

One of the most important ways in which qualitative research is often seen as offering space for the voices of respondents is in the use of direct quotes from participants. Through the use of direct quotes from interview transcripts in written reports, the voices of those interviewed, it is argued, can be heard. Such quotations can take a number of forms, sometimes making use of extensive sections of transcript where both the researcher and researched individual's comments are visible.

The dialogue below illustrates how the interplay between an interviewer and interviewees can be presented. Personal interactions, such as the representation of emotions like laughter, feelings of discomfort, surprise or embarrassment, and physical gestures, in this case nodding, can all be noted within interview transcripts and accompanying notes and reproduced in articles discussing the study's findings.

SM: Do you ever fight with your brothers and sisters about . . .

Carl: Yeah, my sister beats me up.

SM: Your sister beats you up?

Simon: My brother has a go at me, 'cos if he wants to go on the machine I'm on, but he wants to play a different game, like, I don't want to get off my game so he can have a go so we end up having a fight.

SM: Who wins?

Simon: My brother now, 'cos he's a kick boxer! (laughter) So I have to watch it!

SM: What about you, Carl, you said your sister beats you up.

Carl: Yeah, [unclear]

SM: So does she get her own way – does she get to play on it?

Carl: Yeah, probably.

Simon: [laughing at Carl] Does she? [surprised tone]

Carl: Yeah.

Simon: [to Carl] Does she win you if she wants a game?

Carl: Yeah mostly, 'cos she always punches me in the arm and dead-arms me and then mum goes [puts on a high-pitched whine] 'now then stop it you two' and then I always get into trouble.

SM: Does your Mum ever threaten to take it off you?

Carl: [nods]

Simon: She does with me.   (McNamee 1998: 201)

It can also be noted from the extract above how an author must make decisions about whether or not to quote an individual verbatim or whether errors in English should be corrected. Here ' 'cos' has been used instead of 'because' as this is a precise record of the boy's words and reflects his character, including his accent. However, on occasion interviewees may request that their English be corrected in order to save them embarrassment or to help clarify an otherwise confused point. This can be particularly true when working with individuals with learning impairments or mental illnesses (Booth 1996). However, when attempting to clarify a point, researchers must be careful not to change the meaning of a comment altogether, putting words into their respondents' mouths.

Word limits can often mean that these lengthy descriptions, unless key to the message of the paper, give way to more concise and distinctly relevant quotes from interviewees.

> *You feel like something off ... Oh God what's it called ... 'You've been framed!' You feel a*
> *right div.*
> *(Male, visually impaired since birth)     (Butler 1998: 85)*
>
> *I only use it [white stick] when I do need help. And I don't bother with it if I don't*
> *(Female, visually impaired since birth)     (Butler 1998: 94)*

Even with such brief comments, however, points can be made about the positions of the research participants. As illustrated above details of the researched, such as their gender, age or impairment, are given in parentheses after their quotes. Words adding clarity to a quote removed from its full context are added in square brackets. Pauses in conversation are marked by a series of dots. The removal of text from a quote is shown by the use of dots within square brackets and ensures that the reader is aware that the quote has been selective, in order to limit any possible accusations of misquotation.

> *I sometimes hear people talking amongst themselves about me in a sympathetic mode.*
> *Right? Like 'I helped him two years ago.' You know and they'll tell each other, one will tell*
> *the other how [...] and back handed compliments like 'I think it's marvellous what he*
> *does.'*
> *(Male, visually impaired since birth, aged 56–65)     (Butler and Bowlby 1997: 425)*

In some instances explanations of the context of a quote are offered in the text. In the example below an indication of the context in which a comment was made is offered. In other instances details about the environment in which the interview took place, or the political climate at the time the interview took place, may be noted, amongst other factors, as the author feels relevant.

> *When discussing the use of guide canes one young person said: 'I mean if I'm going to a*
> *strange place I take my white stick with me, so that people can see I need help.*
> *I act kind of more helpless than I am to be honest.'*
> *(Female, visually impaired since birth)     (Butler 1998: 93)*

As I suggested above, word limits often make it difficult to incorporate all the material that you would like to about the interviewees in order fully to contextualize what they have to say. One way in which authors often respond to this is to include an appendix, or in a book an opening section, which provides a brief 'pen sketch' of each of the respondents giving fuller details about them (see e.g. Shakespeare *et al.* 1996: 210–11).

The use of the voices of participants as a means to challenge the power relations of the research encounter is not, however, straightforward or unproblematic. Some qualitative research has been the most criticized for its oppressive nature. Stone and Priestley (1996) suggest that participants have on occasion been misrepresented by reports that have quoted them out of context, or misinterpreted their comments. The problems of the researchers' influence on the participants have gone unacknowledged. The temptations for researchers to report only the findings agreeable to their personal political or other agendas can be strong.

Clearly the coding of quotes from interview transcripts in a reflective manner is crucial in terms of how decisions are made about which material will be used and how it represents the arguments that are being developed. It is important that time is taken to note the different perspectives on issues that are raised by the same individuals in different contexts or circumstances. As Crang explained earlier in this book, interview transcripts can be compared for similarities and the different quotes expressing similar points indexed under specific topics and headings. Quotes that relate to more than one topic of interest can be cross-referenced by the analyst for later reflection on the interaction of different phenomena.

Computer packages such as NUD*IST and Ethnograph can be used to search for given phrases in transcripts, which can then be marked and coded according to an allotted heading. Lists of quotes on any given subject, as coded, can then be brought up at the touch of a button and a suitable quote picked to illustrate a given point in a paper. Tape recorders could soon become linked directly to computers with no need for the use of manual transcription services. This could, however, be the start of a slippery slope when it comes to ethical concerns connected to an awareness of positionality and representation of views. Whilst this technology may be time and labour saving, the dangers of such packages, in my opinion, are that they can encourage an author to rely on technology to find common points and patterns in their data rather than familiarizing themselves fully with their data through lengthy reading and rereading of the transcripts. Failure to do so can mean, first, that some contradictory points or rarely referred to points can be lost.

It also means that the author can fail to recognize the broader context in which a quote was made, as the indexed lists of quotes become removed from the full transcript as well as the associated notes made by the interviewer about the interviewees' gestures, behaviour, emotion and so on during the interview. Comments made in a sarcastic manner, for example, may be taken at face value. An analysis of the speakers in a dialogue which will reflect their positionality is required, not simply a carving-up of words.

Conferring with research participants during the writing stages of research can avoid some of these difficulties. Many researchers, including myself, will return a copy of the typed transcript of an interview to an interviewee in order to allow him or her to comment or add to it. Many qualitative researchers equally choose to share the reports that they have written with their interviewees so that they can see how their quotes are used and if they are making the points that they wanted to make. Through this process, you, the researcher and your research participants can open up a dialogue about the analysis and writing of the research – and it may be possible to agree on how differences in interpretation can be written in the text. However, as David Ley and Alison Mountz's discussion about autobiographical writing suggests, there may be times when differences are not reconcilable. One strategy in this case is to ensure that differences of opinion are acknowledged in what is written.

### The researcher's position

Many researchers feel it important to share details about the power relations between the researched and the researcher, not simply the positions of the researched individuals in isolation. The researcher's position is often raised in theoretical writings justifying people's work on different groups of individuals, or relative to methodological or ethical issues, but not connected to writing style (see e.g. Sidaway 1992). There is sometimes limited acknowledgement of positionality, but stating your membership of given social categories has sometimes become as routine, and perhaps meaningless, as the automatic use of statements such as 'depending upon gender, race, class, ethnicity, sexuality, ability or age' have in efforts to acknowledge difference in the populations we study. Limited autobiographical information of this kind has been supplied, I would suggest, according to some sort of recent 'vogue', with little genuine intent to improve the finished article.

Despite feminists', amongst others, recognition of the importance of a researcher's positionality and the need for critical and reflexive engagement with the research process at all stages, autobiographical extracts about the author have been limited in geographical literature (Moss 1999). Feminists in sociology, anthropology and cultural studies have been much more ready, Valentine (1998) argues, to bare their souls (see e.g. Stanley 1992, 1993; Holland and Ramazanoglu 1994; Birch 1998). Whilst some recent work in geography has focused on autobiographical experiences specifically (Chouinard and Grant 1995; Valentine 1998; Moss 1999), most autobiographical anecdotes have been used for one of two purposes.

On the one hand, they offer the reader a sense of warmth and personality. On the other hand, they can infer the author's authority and expertise in a given field. Reginald Golledge (1996: 404) makes this point explicit in his response to comment from Imrie (1996) that his work on visually impaired people implies them to be 'somehow inferior': '... since I am disabled, I would hardly characterize myself so ... From *his* lofty position of ableism, Imrie ... accuses me of denigrating the very group to which I belong and try to represent!'

Rarely are autobiographical notes accompanied by a discussion of their significance and impact on the research findings and presentation in any detail. A noticeable exception is Gleeson (1998: 5):

> *I am not disabled, and, as a white, middle-class male, neither do I directly experience the
> other major types of social discrimination or disadvantage that bear down upon various
> oppressed forms of identity. This fact inevitably limits my ability to understand and explain
> the experience of disablement, in ways that I cannot myself fully appreciate.*

Whatever the aims of brief autobiographical details, once they have been achieved, often at the beginning of a paper or in opening and methodological chapters of theses and dissertations, the personal, Miller (1991) suggests, usually vanishes.

One way in which the author remains visible is through the use of the personal pronoun, 'I'. As Greenfield (1996b: 248) notes, use of 'I'

> *is not simply a matter of taste. It is a question of honesty and the credit that you need
> and deserve ... You are responsible for ensuring that others recognise what you have
> done, what ideas you have had, what theories you have created, what experiments you
> have run, what analyses and interpretations you have made, and what conclusions you
> have reached.   (emphasis in original)*

Greenfield suggests that by using the personal pronoun you are making clear that these are findings that you have created from your specific position.

The twofold solutions to the issues of positionality and self-reflection have been, as outlined above, the raising of the profile of both the researcher and the researched in academic texts, particularly in relation to qualitative research. However, whether the brief references to authors' social positions, or the use of direct quotes from interview transcripts, can justifiably be claimed to clarify the power relations that took place in a research project and clearly pass on to the reader a sense of the position from which a report has been written and hence relates to other work in the field is questionable.

## Conclusion: reflections on theory, ethics and writing

Much work on minority groups, be it in relation to gender, disability, class, sexuality, race, ethnicity or age, has drawn attention to the social construction of people's public images and the differences between such images and their embodied experiences. Whilst an author's aim may be to inform the reader of their relationship to those they work with, stereotypes of their own social categorization may in practice put misleading assumptions in the reader's mind. As I have discussed at length elsewhere (Butler and Bowlby 1997; Butler 1998) impairments, including visual impairments, often stigmatize an individual with negative social constructions of disability, as weak, helpless, mentally inadequate and dependent upon others. If authors therefore declare themselves as impaired or disabled, the same negative images of incompetence and inadequacy could in some instances be prescribed to them by the reader. This may be an extreme scenario, but other concerns about the authors' personal abilities, politics, social and economic status relative to their research subjects can lead to accusations of over-involvement with research subjects, political aims and prior agendas leading to unsurprising conclusions, and so on.

There are clear limitations in previous approaches that gave little attention to positionality. What has not generally been achieved is a reflexive awareness of the impact of the researchers' autobiographies on their writing of others' biographies; the analysis and dissemination of their research. Rather than writing positionality and issues of subjectivity into research literature, authors have left the reader ill-informed of the relationship between the researcher and the researched. For example, points of agreement and disagreement between researcher and researched have not been made clear. Thus, it equally remains unclear as to whether opinions that the author disagrees with have been represented. Readers have been left to make their own conclusions on the representativeness of the arguments made about a research population, based on little more than a stereotype of what the author considers the key social, economic and political characteristics relative to their subjects to be. Different research frameworks have had different methodologies attributed to them. However, I would suggest that the requirements of academic writing have remained relatively narrow and unreflective of these ever broadening methodologies and philosophies. Where an ethical awareness of positionality has been shown throughout a research project, there is a case for this being reflected by the inclusion of autobiographical anecdotes in the written presentation of such projects.

As Rose (1997) argues, reflexivity is not straightforward. Rose criticizes the extent to which we can claim fully to understand our own positionality – pointing to the limits of such knowledge and arguing instead for a more modest writing of ourselves into our research. At the same time, there is concern that 'too much' reflexivity may result in writing that is 'conceited and arrogant' (Greenfield 1996b: 249). Some critics such as Silverman (1997: 239) have called for social scientists to refrain from using the overplayed reflexivity card and return to the issues of aesthetics, noting the primary importance of clarity and reason in writing. This is clearly an issue worthy of debate. 'Clear' writing may mean many different things – as qualitative researchers we want to communicate in ways that readers will engage with; however, we do not want to 'simplify' the complexity of the social world.

At the start of this chapter I did not promise the answers to these questions – and I do not provide them. Rather I hope I have raised an awareness of a need for us all to question when, in what form and to what purpose we declare our 'positionality' in research writings. We are all individuals and as such express our personal viewpoints as part of our daily lives. In an academic research setting, professional training offering an awareness of issues of reflexivity and positionality does not mean that those viewpoints become automatically acknowledged and their impacts upon research recognized. These processes must be worked at and repeatedly returned to throughout a research project, including the writing and dissemination of findings.

The processes of research analysis and the dissemination of findings rely on the thought processes of a positioned individual, however experienced an academic. It must be acknowledged that there is always a need for interpretation and the expression of personal opinion on any particular data set, so that research is more than simply storytelling. There will always be an element of literature that we must take on trust from the author(s). However, positionality is not simply a trendy issue that must be referred to in passing, nor simply a personal one that authors must allow to pinch them from time to time in the privacy of their office when they might question their motives in promoting the findings they do. Positionality is rather, I would suggest, an issue for public debate. Recognition of motivation, differences of position and an awareness of personal reasons for the promotion of particular issues can all be put to valuable use to piece together a more complete picture of society.

## Acknowledgements

I would like to thank Claire and Melanie for all their helpful comments and enthusiastic encouragement.

## Key references

McDowell, L. 1992: Doing gender: feminism, feminists and research methods in human geography. *Transactions of the Institute of British Geographers* 17(4), 399–416.

Rose, G. 1997: Situating knowledges: positionality, reflexivities and other tactics. *Progress in Human Geography* 21(3), 305–20.

Stanley, L. 1992: *The auto/biographical I/eye: the theory and practice of feminist autobiography.* Manchester: Manchester University Press.

## References

Amos, V. and Parmar, P. 1984: Many voices, one chant. *Feminist Review* 3–9.

Andermahr, S., Lovell, T. and Wolkowitz, C. 1997: *A concise glossary of feminist theory.* London: Arnold.

Birch, M. 1998: Re/constructing research narratives: self and sociological identity in alternative settings. In Ribbens, J. and Edwards, R. (eds), *Feminist dilemmas in qualitiative research: public knowledge and private lives.* London: Sage, 171–85.

Booth, T. 1996: Sounds of still voices: issues in the use of narrative methods with people who have learning difficulties. In Barton, L. (ed.), *Disability and society: emerging issues and insights.* London: Longman, 237–55.

Butler, R. 1998: Rehabilitating the images of disabled youths. In Skelton, T. and Valentine, G. (eds), *Cool places: geographies of youth culture.* London: Routledge, 83–100.

—— 2001: A break from the norm: exploring the experiences of queer crip's. In Backett-Milburn, K. and McKie, L. (eds), *Constructing gendered bodies.* London: Macmillan, 224–42.

—— and Bowlby, S. 1997: Bodies and spaces: an exploration of disabled people's use of public space. *Environment and Planning D: Society and Space* 15(4), 411–33.

Carby, H. 1982: White women listen! Black feminism and the boundaries of sisterhood. In Centre for Contemporary Cultural Studies (ed.), *The empire strikes back: race and racism in 1970s Britain*. London: Hutchinson, 212–35.

Chouinard, V. and Grant, A. 1995: On being not even anywhere near 'the project': ways of putting ourselves in the picture. *Antipode* 27(2), 137–66.

Gilbert, N. 1993: Writing about social research. In Gilbert, N. (ed.), *Researching social life*. London: Sage, 328–44.

Gleeson, B. 1998: *Geographies of disability*. London: Routledge.

Golledge, R. G. 1996: A response to Imrie and Gleeson. *Transactions of the Institute of British Geographers* 21(2), 404–11.

Greenfield, T. (ed.) 1996a: *Research methods: guidance for postgraduates*. London: Arnold.

—— 1996b: Writing the thesis. In Greenfield, T. (ed.), *Research methods: guidance for postgraduates*. London: Arnold, 243–52.

Haraway, D. 1988: Situated knowledges: the science question in feminism and the privilege of the partial perspective. *Feminist Studies* 14(3), 575–99.

Holland, J. and Ramazanoglu, C. 1994: Coming to conclusions: power and interpretation in researching young women's sexuality. In Maynard, M. and Purvis, J. (eds), *Researching women's lives from a feminist perspective*. London: Taylor & Francis, 125–48.

Imrie, R. 1996: Ableist geographies, disablist spaces: towards a reconstruction of Golledge's 'geography and the disabled'. *Transactions of the Institute of British Geographers* NS 21(2,) 397–403.

Levy, P. 1996: Presenting your research: reports and talks. In Greenfield, T. (ed.), *Research methods: guidance for postgraduates*. London: Arnold, 253–68.

McDowell, L. 1992: Doing gender: feminism, feminists and research methods in human geography. *Transactions of the Institute of British Geographers* NS 17(4), 399–416.

McNamee, S. 1998: Youth, gender and video games: power and control in the house. In Skelton, T. and Valentine, G. (eds), *Cool places: geographies of youth culture*. London: Routledge, 195–206.

Miller, N. K. 1991: *Getting personal: feminist occasions and other autobiographical acts*. London: Routledge.

Morris, J. 1992: Personal and political: a feminist perspective on researching physical disability. *Disability, Handicap and Society* 7(2), 157–66.

Moss, P. 1999: A sojourn into the autobiographical: researching chronic illness. In Butler, R. and Parr, H. (eds), *Mind and body spaces: geographies of illness, impairment and disability*. London: Routledge, 155–66.

Oliver, M. 1994: Re-defining disability: a challenge to research. In Swain, F., Finkelstein, V., French, S. and Oliver, M. (eds), *Disabling barriers – enabling environments*. London: Sage, 61–7.

Phillips, E. M. and Pugh, D. S. 1987: *How to get a Ph.D.: a handbook for students and their supervisors*. Milton Keynes: Open University Press.

Rose, G. 1997: Situating knowledges: positionality, reflexivities and other tactics. *Progress in Human Geography* 21(3), 305–20.

Shakespeare, T., Gillespie-Sells, K. and Davies, D. 1996: *The sexual politics of disability*. London: Cassell.

Sidaway, J. D. 1992: In other words: on the politics of research by 'First World' geographers in the 'Third World'. *Area* 24(4), 403–8.

Silverman, D. 1997: Towards an aesthetics of research. In Silverman, D. (ed.), *Qualitative research: theory, method and practice*. London: Sage, 239–53.

Skelton, T. and Valentine, G. (eds) 1998: *Cool Places: geographies of youth culture*. London: Routledge.

Stanley, L. 1992: *The auto/biographical I/eye: the theory and practice of feminist autobiography*. Manchester: Manchester University Press.

—— 1993: On autobiographies in sociology. *Sociology* 27(1), 41–52.

Stone, E. and Priestley, M. 1996: Parasites, pawns and partners: disability research and the role of non-disabled researchers. *British Journal of Sociology* 47(4), 699–716.

Valentine, G. 1998: 'Sticks and stones may break my bones': a personal geography of harassment. *Antipode* 30(4), 305–32.

# PART 7

## Vignettes

In this final part of the book five geography students discuss the research projects that they undertook as part of their geography degrees. Each short vignette discusses the methodological choices that the individuals made about how to organize their research and each author also reflects on the research process as a whole, considering both the strengths and weaknesses of the approaches she used. These vignettes offer an insight into the ways in which undergraduate researchers, often embarking on a research project for the first time, are using qualitative methodologies. These accounts show how the authors were involved in combining different research methods and also highlight the ways in which research contacts were made and interviewees or participants recruited. Finally, each account offers a helpful reflection about how the researchers themselves considered their own position in relation to the research that they undertook.

# Vignette I

**Schooling and striving for success: an investigation into the aspirations of girls attending high school in rural south India**

Katharine Moss

My research began in December 1998, when I stumbled across the Bhagavatula Charitable Trust (BCT) on the Internet. Within one week I was in India, working in a predominantly rural area of the state of Andhra Pradesh. Casual conversation with schoolgirls revealed that they wanted to study more, and obtain paid employment, and that their general attitude towards the future was very positive. On my return, I was therefore surprised to read that recent case studies, particularly in South Asia, have cast doubt on the precise gains of schooling for women, especially the assumption that schooling enables women to obtain paid employment (Bourque and Conway 1995; Basu and Jeffery 1996; Jeffery and Jeffery 1997; Bunwaree and Heward 1999). On reflection I realized I had met few working women, which led me to consider the possibility of the existence of a disjuncture between girls' aspirations and achievements.

I decided, therefore, to return to the same area to examine schoolgirls' aspirations and the likelihood they would fulfil them. To explore aspirations, I used a mixture of 'quantitative' and 'qualitative' methods. I drew a sample of 117 girls from three different high schools. Two of the schools were government run; one, Dimili High School, was mixed, whereas the other, Yellamanchili High School, was single sex. The third school was mixed and was established by BCT. By comparing the results from each of the schools I hoped to ascertain whether school type had any influence on the girls' aspirations.

First all 117 girls completed structured questionnaires written in Telugu (the local language). In these I asked them to provide details of their schooling, what they wished to do after Class 10 (the last year in all the local high schools), together with their material and social aspirations. There was insufficient time to do a pilot study but I sought extensive advice from Indian academics, government and UN officials, together with those living in the area concerned. In Yellamanchili 39 of the 154 girls in Class 10 were selected by the headmaster, drawing equally from the three income brackets. In BCT and Dimili all the Class 10 girls were interviewed, together with younger girls to gain a sufficient sample, in total 46 from BCT and 32 from Dimili. This unfortunately led to a significant difference in the average age of the girls interviewed in the three schools. It was also not ideal that the Yellamanchili headmaster chose the girls from his school, but I had no option, as he would not allow me to see the registers and select my own sample.

Once translated into English, these data were analysed using descriptive statistics and chi-square tests (95 per cent significance level) to determine if aspirations varied significantly between the three schools. I then used these quantitative data as a sampling frame for individual in-depth interviews with selected girls with older sisters. These interviews were conducted with girls individually with the aid of an interpreter. In these interviews I enquired about the older sisters' school backgrounds, what they wanted to do and had done since leaving school, and used this information to indicate whether it was likely that the girls currently in school would achieve their aspirations. I also discussed the schoolgirls' opinions on their schooling, including whether they would like more vocational training. During the interviews I tried to make the girls feel as comfortable and unhindered as possible by talking to them on their own away from distractions and recording the interviews to avoid extensive note taking. This also enabled me to have a subsequent independent translation of the discussions. On occasions, however, several girls were very shy and seemed unable to form their own opinions.

Over the three months I was in India I also collected information from Indian academics, the Indian government, the United Nations and other non-governmental organizations. During this time I attempted to immerse myself in Indian culture as fully as possible, even learning some Telugu to understand what it is like to be a woman in India. From this I hoped to understand what would constitute 'success' in the eyes of an Indian schoolgirl. This is intentionally subjective, for success does not necessarily mean obtaining paid employment or having a set number of children. For too long women's aspirations have been assumed using entirely Western ideas, without considering different cultural contexts and perceptions of inequality (Basu 1996). I believe that it is only through detailed case studies, asking girls about their aspirations, that they can be appropriately assisted to achieve what for them constitutes success.

As with any fieldwork I experienced a number of difficulties, from sickness and coping with 'Indian timing', to my interpreter pulling out the day I was flying out to India. Eventually, however, I managed to collect sufficient data for my purposes, discovering overall that schooling provides girls with *aspirations*, such as obtaining paid employment, which are consistent with the benefits of schooling assumed by organizations like the World Bank. There was, however, little evidence to suggest that these aspirations would be fulfilled, suggesting the actual *actions* of women are mediated by cultural attitudes. Above all, however, the girls themselves were tremendously optimistic about their chances of success and there was a general feeling amongst them that things were changing, with the situation for women becoming generally easier. I therefore concluded by arguing that, irrespective of background, schooling is extremely important in helping to overcome disparities in life chances, giving girls aspirations and ideas about how things can be different. Schooling does not, however, automatically lead to change and must be viewed as only part of the process through which women can obtain control over their lives. To ensure aspirations are realized, policies are needed that place schooling in a wider context. The influence of societal expectations, especially the role men ascribe to women, must be acknowledged. Challenging such well-established cultural norms is likely to take a long time, but my research demonstrated that integrating schooling with community action can cause significant changes. Above all, women must be valued irrespective of whether they work and all girls must be helped wherever possible to achieve what for *them* is success.

A major problem with this research is the use of older sisters as analogues for the girls' future lives. The influence of birth order on life chances of siblings (Satish 1995), combined with the feeling among the girls that change was occurring, made them particularly unreliable. Ideally, therefore, the same girls should be reinterviewed in several years, comparing their aspirations with what they have *actually* done. Another problem was the language barrier, which prohibited some data collection and it was difficult to convince the translator not to summarize in-depth answers.

The research also clearly highlighted the importance of geography and the fact that data cannot be generalized to represent the case of all Indian schoolgirls. Similar studies should be conducted in other situations in India, north and south, rural and urban, together with different caste and class groups, to determine whether the disjuncture I discovered between aspirations and achievements is widespread. Parents, boys and girls not in school should also be interviewed to give a clearer idea of the factors influencing the realization of ambitions and whether it is indeed the structures of society that hinder girls' chances of success.

## References

Basu, A. 1996: Girls' schooling, autonomy and fertility change: what do these words mean in South Asia? In Basu, A. and Jeffery, R. (eds), *Girls' schooling, women's autonomy and fertility change in South Asia.* London: Sage, 15–47.

—— and Jeffery, R. (eds) 1996: *Girls' schooling, women's autonomy and fertility change in South Asia.* London: Sage.

Bourque, S. C. and Conway, J. K. (eds) 1995: *The politics of women's education: perspectives from Asia, Africa and Latin America.* Michigan: University of Michigan Press.

Bunwaree, S. and Heward, C. (eds) 1999: *Gender, education and development: beyond access to empowerment.* London: Zed Books.

Jeffery, P. and Jeffery, R. 1997: *Population, gender and politics: demographic change in rural North India.* Cambridge: Cambridge University Press.

Satish K. 1995: Gender issues and quality of work life: a case study of the formal labour market in Delhi. *Labour and Development* 1(1), 43–4.

# Vignette 2
## Being Asian, being British and being into black music: the influence of black music on hybrid identities of young South Asian women in London
## Sunita Ram

Previous studies of young Asian women have tended to focus on more 'traditional' and religious aspects of their lives. My dissertation highlights the ways in which black music (R&B, hip hop) has had a significant impact on youth culture in Britain and looks at its influence on the identity construction of young Asian women. The dissertation reveals how the participants negotiated their inherited Asian cultures and their identities as British Asians living in multi-ethnic environments. All of the participants saw their identities as hybrid, combining Asian, British and black cultural aspects.

Since an appreciation of black music takes young Asian women into the social spheres of black music clubs, the dissertation also focused on the politics of the club space as well as examining some of the inter-racial relations that were constructed in these social spaces. For example, respondents talked about how they negotiated their identities in relation to young Caribbean women in these spaces and the politics of 'looking black' in terms of negotiating fashion. The dissertation sought to expand our understanding of young Asian women by going beyond traditional representations and stressing the multiplicity of different hybrid identifications possible, particularly within a multi-ethnic city like London.

Given my own involvement in this topic as an Asian woman who enjoys black music and visits black clubs, as well as working as a DJ (see Fig. V2.1), I was able to conduct this research through a combination of ethnographic research and semi-structured interviews. I conducted 10 in-depth interviews. Through discussing my ideas for my dissertations with friends, I was put in touch with most of the participants, so they were all either acquaintances of mine or discovered through 'snowballing' of existing contacts. I did recruit two interviewees through approaching them at clubs – although potential participants at clubs were sceptical when they were approached, as they were not convinced that a dissertation could be focused on such an exciting topic! In selecting the participants I felt that it was most important that they had a strong liking for black music, experience of clubbing and close non-Asian friends. This meant that I did not target respondents from particular ethnic backgrounds (although all of them grew up in Britain),

**Fig. V2.1** *Sunita Ram as DJ*

and consequently the Asian backgrounds of the respondents vary considerably. Instead, I made their strong interest in black music and black culture the most important criterion. My respondents also included an MTV producer and presenter. I recruited her by sending her an 'eye-catching' fax, which proved to be successful in attracting her attention, and she invited me to interview her at the MTV studios. I also interviewed JC001, a male Asian rapper.

I conducted all of the interviews in a university union meeting room, which was a convenient and neutral space. I made an interview schedule before the interviews, which set out the themes of the interview. I began by focusing on the interviewees' experiences of clubbing – where they go, what kind of music they like, when they first became interested in black music. I then deepened the focus by asking about identity and clubbing – did they feel that, being Asian, they were treated any differently in any clubs, was 'race' relevant? How did their parents or Asian friends feel about their interest in black music/clubs? The final part of the interview focused more on how respondents defined their own identities and how they responded to issues such as 'multicultural' or 'between two cultures' or Asian 'sell-outs' or appropriation of black culture. In practice the interviews were very conversational, and sometimes the participant and myself strayed away from the initial questions. Being a young Asian women myself meant that I was 'on the same level' as the interviewees as common experiences arose. It could be argued that a non-Asian interviewer might have gained different responses. When I asked the respondents to reflect on the interview process, they had very much enjoyed it, pointing out: 'the topics prove how many of us are unrepresentative of people's stereotypes'; 'there's more to our lives than people think'; 'it's been wicked to talk about how important music is to me, Asian girls just aren't seen as music fanatics or DJs'.

When I visited nightclubs as part of the ethnographic research, I kept a research diary after my visit. This included details about the kind of music played, what kinds of people were there, what kinds of clothes they were wearing, the visuals shown in the club as well as any flyers, and so on, that were distributed. I also kept a record of any of the interactions that I noticed in the club amongst groups of people – for example, whether friendship groups were ethnically mixed or segregated or the kinds of 'competition' between groups of girls that were taking place for male attention. In writing up the dissertation I tried to incorporate the interview material – as quotes – with some of the ethnographic material that was written through from the research diary in the first person. The way I chose to do this was to write the main narrative within the chapter, interspersing this with quotes from the respondents. I would then make reference to the ethnographic material and include this as an extract from my research diary but have this as a piece of boxed text, which could be referred to in the main text. I also included lots of visual material in the dissertation, including photographs from the clubs, which illustrated some of the points that I wanted to make, as well as flyers and other visual material from magazines. When I handed in my final dissertation, I also included a tape that illustrated some of the music discussed by the respondents.

Reflecting on the overall research for my dissertation, I would have liked to deepen it by incorporating more ethnographic research to provide greater context for the interviews. I could have done this by making more follow-up visits to the same clubs as well as other clubs mentioned by the interviewees. I could also have strengthened the focus on the club itself by interviewing the club promoters and talking to other clubbers about the presence of Asian women.

# Vignette 3
## Househusbands: the socio-spatial construction of male gender identity
Annabelle Aish

My dissertation explored the socio-spatial construction of male gender identity, focusing on men who had adopted the conventionally female role of homemaker, becoming 'househusbands'. Men's changing relationship to the domestic sphere has been well documented; however, there have been relatively few studies on fathers who carry out housework and childcare on a full-time basis. Most men remain 'breadwinners' who locate their primary identity in the workplace and play a secondary role in the home. Yet, as women increasingly command similar 'breadwinning' salaries, some husbands argue that it is more logical for them to be 'homemakers' rather than 'providers'.

This study examined the ways in which working in the traditionally female sphere of the house has affected househusbands' conception of themselves *as men*. I found that, by perceiving homemaking as a 'job' that demands the mastering of certain (feminine) skills, these men have renegotiated their role as a worthwhile and fulfilling challenge. For example, some of the men interviewed talked about the ways in which this experience gave them new skills or abilities they could bring back to the paid workplace. As such, these househusbands were fundamentally contesting the traditional notion of masculinity by maintaining that it is possible to be both 'a man' and 'a good father' in this feminine role. Furthermore, by becoming homemakers, they were challenging the female nature of the home and actively loosening the rigid dualisms of man/woman, reason/non-reason, public/private situated at the work/home binary.

The research was conducted through semi-structured interviews with a group of 12 men. They were all heterosexual men in long-term/married relationships and acted as primary carers to children under the age of 16. I contacted these men through an organization called 'Househusbands Link', the only parenting organization specifically for men who are primary carers for their children. 'James', who ran this organization, became my 'gatekeeper' and gave me the names and addresses of his members as well as his open support for my research. I contacted those members who were living reasonably close to London (where I was based) and invited those who were willing to participate in a group interview. I conducted two separate group interviews, one with four participants, one with three.

These group interviews were useful in exploring interactions between the participants. They were informal discussions based around a checklist of topics, which included feelings about leaving their former workplace and taking on the traditionally female role of homemaker, as well as their everyday experiences of housework and childcare. I was helped in constructing this checklist by using the 'Househusband Link' newsletter (written by the group), which informed me of issues that might be of significance. After the group interviews I decided that I wanted to discuss some issues in further depth and so chose also to conduct one-to-one interviews. This also enabled me to include some additional interviewees. I held these interviews in the fathers' homes, which was most convenient for them – and put them at their ease. I was able to rope in a friend to help take care of the children while I conducted the interviews. I also conducted three telephone interviews with men living too far away to visit. Despite being concerned about these interviews, they proved to be more productive than I had imagined, as the interviewees were equally forthcoming. I also handed out a short questionnaire to each participant, which provided me with basic information about the socio-economic and educational status of these men and their partners. However, the in-depth interviews took priority and seemed to be the most appropriate way of drawing out ideas concerning identity, private feelings and personal understandings.

The personal nature of the conversations made it appropriate to guarantee these men's anonymity and therefore I changed their names and those of their wives and partners. After I had transcribed each interview, recurring themes were drawn out through 'open coding' and I eventually developed a set of themes and subthemes and finally a set of generalizations that explained the most significant findings. Of course these generalizations remain a partial and 'positioned' interpretation and only an abstraction of the complex web of interrelated and contradictory feelings expressed. Thinking about how the 'social encounter' of the interview shaped my research, I felt that the atmosphere in the interview was open and friendly. Although I cannot tell how the men's opinions of me as a young, female researcher influenced what they chose to tell me, I did not sense any reticence or lack of honesty. In terms of my own positionality, I am not a parent nor have I been in long-term employment like the househusbands interviewed. Perhaps most importantly, I am not male. I cannot pretend to have personally experienced the complexities of masculine identity or how it feels to go against the grain and assume 'a woman's place'.

Reflecting on my dissertation, I realize that there are inevitable limitations to the research. It was disappointing not to interview all the fathers face to face because of their location. I was also left with conflicting feelings about the group interviews. While some of the men may have been more open in a group environment (with people they already knew), others may have felt obliged to give answers that they thought were acceptable to the group. As my sample was unavoidably small, I cannot begin to assume my study incorporates all fathers' experiences of househusbandry. In particular, my interviewees were mostly middle-class men, all belonging to 'Househusbands Link', an organization whose 153 members are mostly from southern England. I have no doubt that there are some fathers who do not consciously identify themselves as 'househusbands' or are unwilling to join what they perceive as a middle-class 'men's group'. If this study were expanded, it would be interesting to see to what extent men's experiences were due, at least in part, to where they lived. There are thus some questions that remain unanswered in my research. In many ways, this was a reflection of the limited number of voices I could include. Fundamentally, I did not include the wives and partners of the men interviewed and so cannot provide any conclusions as to how they understood their non-traditional family structure. Overall, however, I am pleased with the way in which the research went. The men I interviewed were also interested in reading about the research and I wrote an account of the research for the 'Househusbands Link' newsletter, as well as giving 'James' a copy of the results and the conclusions.

# Vignette 4

## Representations of 'race' in *Goodness Gracious Me*: comparing British South Asian and white British responses to the programme

## Sarah Corke

*Goodness Gracious Me* is a BBC comedy written by a team of four British Asian actors. It is innovative in taking a satirical look at British and Asian culture and their interaction. It is a unique example of representing hybrid cultures in a white-dominated market. It has received critical acclaim and won the Commission for Racial Equality Media award. Given that different groups of people 'read' media programmes differently, based, in part, on their own 'race', class and gender positions, I focused my research on how 'Asian' and 'British' groups responded to the representations of British and Asian people in the programmes.

Racial terminology is often conflicting and contested. It is largely understood by geographers that the term 'race' has no biological basis and is, in fact, a social construction. Since *Goodness Gracious Me* is about how British and Asian cultural ideas and behaviours intersect, I decided to use the way in which the programme constructs these two groups as the basis for determining my research group samples. Thus, one group of respondents, defined as 'South Asian', were those who had South Asian heritage although they had always lived in Britain. The group of respondents defined as 'white British' were those who had an English, Scottish or Welsh heritage and identify as 'white'. The other criterion for recruitment was that the respondents were regular watchers of the programme.

The research methodology revolved around group discussions with two focus groups from each 'ethnic group'. I chose focus groups as my methodology because they best recreated how people discuss television shows in real life. The group discussion gave participants the opportunity to consider, challenge or take on board other people's readings of the representations. It was interesting and important to see which opinions changed when challenged by other group members, and which opinions were shared by all in the group.

Each group contained six people between the ages of 16 and 24. I chose this age group because I wanted to recruit participants who were my peers and thus more willing to talk to me. I contacted Asian and white friends and asked them to snowball new recruits who would be interested in the project. This technique avoided the possibility of the group being overly familiar, which might have produced similar points of view, but ensured all participants knew at least one person in the group. I held all the groups in my house, which provided surroundings most were familiar with. I started each group with informal introductions and allowed the group to talk or catch up before I launched into the discussion. This was surprisingly helpful, as it made people more relaxed about putting forward their point of view during the session.

Most people found the focus groups an unusual way to conduct research. Some participants asked for the correct answers beforehand, because they were worried they would not say the 'right' thing. I emphasized that the discussion was about their responses to the representations in the programme and that there was no 'right' or 'wrong' answer. Before the discussion I showed the group an episode of *Goodness Gracious Me* (series three). I used this as a stimulus for the discussion but did not limit the discussion to this one episode. I began each group with a starter question: 'How do you think Asian and British people were represented in the show?' This was a deliberately vague question, which allowed participants to begin by discussing elements that were important to them. The answer of the first respondent was soon picked up by another person and, if not, I asked the next person to comment on the show. Unlike structured interviews, focus-group members are empowered by being able to decide the itinerary of the discussion. It was just as important for me to acknowledge the themes that some groups left out of their discussion as well as the ones they talked about.

Having never moderated a focus group before I was scared that I would not be able to keep the group on track. I prepared a guidesheet containing questions that I could ask if the conversation died or strayed. I did, however, find that allowing my groups to wander off track meant they were more willing to return to the topic without getting bored. This and other techniques for moderating the group became apparent to me, the more focus groups I conducted. The important thing for me was learning when to contribute and when not to. The participants often directed questions to me, about the programme or my ethnicity, which I was unsure whether to answer or not. I soon learnt that, if I did not answer, the group dynamic was disrupted and participants immediately became aware of my position as the researcher. I offered my opinion and found ways to redirect the question to another person, often to a participant who was more inhibited. I found quieter participants often had things to say but not the opportunity to say them. By providing this window for them I ensured the louder participants did not dominate. I allowed the groups to run as long as the members remained interested, which ranged between one and two hours. It soon became evident when people became tired and I ended each session by asking if anyone had anything they wished to add to ensure everyone felt they had been heard.

I am very happy with how the focus groups went. What is more, I enjoyed taking part in them. The issue of 'race' and representations of 'race' sparked off much passionate debate. Many of the respondents told me that they had never thought about their opinions on the matter let alone been asked about them. I often had people discussing afterthoughts from the group with me. I had originally planned to conduct repeat sessions with my four groups in order to discuss such afterthoughts and create a stronger group dynamic. However, I found that, although people were willing to do another session, varying timetables made it impossible to organize. I was worried that without the repeat sessions I would not have enough data, but after analysing pages of transcripts I have realized that focus groups provide you with plenty of data-rich information.

# Vignette 5

## The retail workplace geography of gender, space and power dynamics: a case study of a camping business in south-east England

## Philippa Capps

The aim of my dissertation was to explore the retail workplace geography of gender, space and power dynamics using the case study of a camping business in south-east England. Workplace geographies of the retail industry have, hitherto, received little attention and my case study aids in developing this field of research. Interactions between co-workers themselves, and between staff and customers, were investigated, concentrating on gender, space and power dynamics. Using qualitative techniques, the study examines how the construction of this gendered organization, the identity politics of individuals, and the performance of retail work all influence, and are influenced by, lateral and vertical relations. It reveals how important all of these issues are in the workplace, and their consequences in the changing patterns of interactions between men and women.

The qualitative techniques used included covert and overt participant observation, focus groups, semi-structured interviews and co-worker surveys. The main emphasis was on participant observation, but, in order to enhance the validity and reflexivity of the research, I used these triangulation procedures to complement the study. In this way, I could support my own 'situated knowledge' (Haraway 1990) with the participants' views and actions.

Both covert and overt participant observation helped to elucidate many facets of the structure and agency of this retail workplace and the complexity of human relations.

During the months of July and August 1999, I became an ethnographer as well as a worker at Westbury's (pseudonym). As a worker myself, I used the covert method, so that nobody treated me or behaved differently (Cook 1997), and so that I could try 'to achieve an empathic knowledge of the state of mind of the actor(s) to reconstruct social phenomena' (Evans 1988: 199). Daily events that I perceived to be important to the study were recorded in a diary. I wrote brief notes at work, which were elaborated upon later in the day, or, if the diary was not to hand, comments were written on scraps of paper. Recording trigger words helped me in remembering the event accurately, and interesting quotes of various colleagues were written down as soon as possible. My lunchtimes were often spent in quiet locations watching events, which helped to make the observation as discreet as possible. However, I had to take care as to where I left my materials. One funny incident did take place when one of my colleagues found my scribbles which read: 'Old men, hiking equipment, boys' club'. My younger colleague wondered what I was up to, but fortunately she thought that I was going slightly mad! I managed to wriggle out of the situation by simply saying that I had been bored, and was doodling as I watched the men. I took more care in the future! I was lucky that it was not something that I had written about one of my colleagues, because, as they were unaware of the research, they might have taken this personally. I found that, when conducting participant observation, it is important to monitor your own actions, as well as observing other people. This is because my colleagues, especially the younger ones, could easily have mirrored my views and actions, so my views had to stay very much covert.

In the last two weeks of the study, a letter was given to each participant explaining what I had been doing, and assuring them of compete confidentiality and the use of pseudonyms in the final report. They all agreed to the covert participant observation that I had been conducting. They signed consent forms, completed semi-structured surveys and participated in focus groups. These results aided in supporting my arguments in the report, and made the covert participant observation very relevant. This is because, once the participant observation became overt, the subjects' attitudes and actions towards me did change considerably. They were more wary with what they said and approached me making comments such as: 'I've just seen something that you could use in your project. Let me tell you what happened.' These changes were very relevant in the case-study analysis, and the use of triangulation procedures helped to strengthen the validity of the research, and the contrasts between the results of the covert and overt participant observation were extremely enlightening.

It is sometimes argued that covert analysis is unethical, but I did obtain the permission of the Managing Director before commencing, and I also showed the final report to all the consenting participants to ensure that there were no objections. It has also been argued that over-familiarity with the research setting may have resulted in too much participation at the expense of observation. However, in my opinion, the length of time that I had been working for the company, and the intimate relationship that I had developed with the management and my co-workers, put me in an excellent position to understand and research the workings of the company on a day-to-day basis.

When I was analysing the study, the data were separated into the relevant categories, such as examples of embodiment labour, and identity politics of individuals, so that the pitfall of being swamped by vast amounts of information and not knowing where to begin was avoided. It was also essential to step back from the study, as it is easy to become immersed in the research when conducting participant observation.

In the final report, I found it extremely important to insert snippets and quotations to justify my arguments, and it also aided in bringing the study to life. The intentional use of the first person also increased the vitality of the work. In my opinion, participant observation is a very useful tool to create feeling in the research, and it helps readers to gain a sense of understanding of the subjects being studied. It is a time-consuming and challenging methodological procedure, but the results make it all worthwhile.

## References

Cook, I. 1997: Participant observation. In Flowerdew, R. and Martin, D. (eds), *Methods in human geography: a guide for students doing a research project*. London: Longman, 127–50.

Evans, M. 1988: Participant observation: the researcher as research tool. In Eyles, J. and Smith, D. M. (eds), *Qualitative methods in human geography*. Cambridge: Polity, 118–35.

Haraway, D. 1990: *Simians, cyborgs and women: the reinvention of nature*. London: Free Association.

# Index